# Studies in Computational Intelligence

Volume 836



**Series Editor**

Janusz Kacprzyk, Polish Academy of Sciences, Warsaw, Poland

The series "Studies in Computational Intelligence" (SCI) publishes new developments and advances in the various areas of computational intelligence—quickly and with a high quality. The intent is to cover the theory, applications, and design methods of computational intelligence, as embedded in the fields of engineering, computer science, physics and life sciences, as well as the methodologies behind them. The series contains monographs, lecture notes and edited volumes in computational intelligence spanning the areas of neural networks, connectionist systems, genetic algorithms, evolutionary computation, artificial intelligence, cellular automata, self-organizing systems, soft computing, fuzzy systems, and hybrid intelligent systems. Of particular value to both the contributors and the readership are the short publication timeframe and the world-wide distribution, which enable both wide and rapid dissemination of research output.

The books of this series are submitted to indexing to Web of Science, EI-Compendex, DBLP, SCOPUS, Google Scholar and Springerlink.

More information about this series at http://www.springer.com/series/7092

Aboul Ella Hassanien · Ashraf Darwish ·
Hesham El-Askary
Editors

# Machine Learning and Data Mining in Aerospace Technology

Springer

*Editors*
Aboul Ella Hassanien
Faculty of Computers and Artificial Intelligence, Information Technology Department
Cairo University
Cairo, Egypt

Ashraf Darwish
Faculty of Science
Helwan University
Cairo, Egypt

Hesham El-Askary
Center of Excellence in Earth Systems Modeling and Observations, Schmid College of Science and Technology
Chapman University
Orange, CA, USA

Department of Environmental Sciences
Faculty of Science
Alexandria University
Alexandria, Egypt

ISSN 1860-949X ISSN 1860-9503 (electronic)
Studies in Computational Intelligence
ISBN 978-3-030-20214-9 ISBN 978-3-030-20212-5 (eBook)
https://doi.org/10.1007/978-3-030-20212-5



This Springer imprint is published by the registered company Springer Nature Switzerland AG
The registered company address is: Gewerbestrasse 11, 6330 Cham, Switzerland

# Preface

Space technology has become an integral part of critical infrastructures and key elements for great power. Space telemetry data contain a wealth of information about the system behavior of artificial satellites. Recent developments in data mining techniques for anomaly detection, fault detection and prediction make it possible to examine this data and extract embedded information to produce advanced system health monitoring applications. Determining the health state of artificial satellite systems using traditional methods is becoming more difficult as thousands of sensor values of different subsystem and component interactions grow. Due to the inherent properties and high complexity of the telemetry data of space systems, conventional methods are not sufficient for this task, the major reason for the difficulties in conventional methods (e.g., limit checking, expert systems, and model-based diagnosis), they are heavily dependent on a priori knowledge on the system behavior for each space system. Moreover, there still exist a number of anomalies or their symptoms, which cannot be detected just by monitoring whether sensor values are between upper and lower limits. In other words, some class of anomalies occurs without violating the limits on the variables.

Data mining is a multidisciplinary field that includes machine learning, artificial intelligence, database technology, pattern recognition, statistics, expert systems, knowledge acquisition, and data visualization. Space missions addressing different science questions related to the earth's varying spheres are on the rise. Hence, monitoring artificial satellite's health and functioning is of great importance using the wealth of the streamed telemetry data received at the ground control units. Owed to the large volume of telemetry data collected, either in real time or saved mode, during the mission's lifespan, data mining algorithms have been applied recently for data handling. Such algorithms are used for analyzing the satellite's telemetry data for anomalous behavior detection or for potential failures prediction. These failures include altitude determination, subsystem control, power, and other parameters for onboard subsystems.

Telemetry, tracking, and control subsystem of an artificial satellite is the brain of the artificial satellites and all spacecraft, which provides a connection between the satellite itself and the facilities on the ground. One of the main functions of this subsystem is to ensure the satellite performs correctly. Any fault in telemetry, tracking, and control subsystem causes loss of the control over the satellite or the spacecraft.

Telemetry is the link from satellite to ground station, non-stationary time series dataset contains thousands of sensor measurements from various subsystems, which contain the wealth information related to the health and status of the entire satellite and all its subsystems, space environment, and others, which reflect the operational status and payload of satellites. Telemetry data contains thousands of sensor outputs from multiple different subsystems and each one of these subsystems brings up to thousands of records every day representing health, status, and mode of each one, besides thousands of the environmental changes and attitude of the satellite measurements. Telemetry data has some important characteristics such as high dimensions, heterogeneity, multi-modality, and missing data.

This book explores the concepts, algorithms, and techniques of data mining in analyzing telemetry data of satellites for health monitoring. It presents an experimental implementation of telemetry data processing to obtain hidden events using different data mining techniques. In addition, the book aims to provide the readers, scholars, and researchers with basic knowledge of satellite monitoring and data mining for anomaly detection and prediction targets.

Editors of this book would like to express their gratitude and thanks for all participants and authors of this book.

Cairo, Egypt — Aboul Ella Hassanien
Cairo, Egypt — Ashraf Darwish
Orange, USA — Hesham El-Askary

# Contents

# Part I
# Health Monitoring of Artificial Satellites

# Tensor-Based Anomaly Detection for Satellite Telemetry Data

**Alaa H. Ramadan, Aboul Ella Hassanien, Hesham A. Hefny and Lamiaa F. Ibrahim**

**Abstract** Satellites is the bird's-eyes that enable us to view massive areas of earth at the same time, satellites can gather more data, more quickly, than tools on the ground. Satellites also can view into space better than telescopes at earth's surface. Development of such artificial satellites which composed of many subsystem, requires a lot of time and money that any deadly failure is unacceptable, the satellite operating in a remote environment, so it is practically very hard or impossible to repair it once a severe failure occurs, so detecting the anomalies of the subsystems measurement values (Telemetry data) is the first step in satellite failure protection and early warning. Traditional spectral-based methods like PCA is traditional for detecting anomalies in a variety of domains and problems. However, if the collected data contains tensor (multiway) structure, for example space-time-measurements values, such as the satellites subsystems measurement, some significant anomalies may stay hidden with these traditional methods. Tensor-based anomaly detection (TAD) applied in a variety set of disciplines over the recent years, although it is not recognized yet as an official category of anomaly detection techniques. This work target to highlight the candidate of tensor-based technique as a new approach for identification and detection of abnormalities and dud in the satellite telemetry data.

**Keywords** Tensor · Satellite telemetry data · Anomaly detection

## 1 Introduction

Launching satellite to the space is a very expensive costly task. The two Egyptian satellites cost around $60 million,[1] so keeping the satellite in the orbit is a national mission. Monitoring the satellite through entire mission life handle by the received

[1] https://en.wikipedia.org/wiki/Category:Satellites_of_Egypt.

A. H. Ramadan (✉) · A. E. Hassanien · H. A. Hefny · L. F. Ibrahim
Institute of Statistical Studies and Research, Cairo University, Cairo, Egypt
e-mail: alaa_hedib@pg.cu.edu.eg

 
A. E. Hassanien et al. (eds.), *Machine Learning and Data Mining in Aerospace Technology*, Studies in Computational Intelligence 836,
https://doi.org/10.1007/978-3-030-20212-5_1

telemetry data from its subsystems, these data related to the measurements concern magnitudes such as temperatures, voltages, currents. The real-time detection of malicious or abnormal behaviors is of critical importance to the safety of the satellite. Detecting the anomalies in the subsystems measurement values is the first step in satellite failure protection and early warning. Furthermore, anomaly detection is the core function on prognostics and health management (PHM) that applied much into the space engineering. Classic spectral-based techniques like PCA are common for detect anomalies in a multiple types of problems and areas. However, when the data set contains tensor (multiway) structure for example time-space- measurements, as the satellites subsystems measurement, somewhat meaningful anomalies may stay invisible with these traditional methods. Tensors are generalizations of vectors (first-order tensors) and matrices (second-order tensors) to arrays of higher orders. In recent years, some methods based on tensor description have been proposed and perform well for anomaly detection (AD), in addition many researches about tensor-based anomaly detection (TAD) increasing. Moreover, there are many methods developed in various disciplines from environmental monitoring, chemometrics, and social networks to data mining and signal processing. Tensor-based anomaly detection (TAD) applied in a set of disciplines over the recent years, although it is not recognized yet as an official category of anomaly detection. This work target to highlight the candidate of tensor-based technique as a new approach for identification and detection of abnormalities and dud in the satellite telemetry data.

## 2 Satellite Telemetry Data Anomalies Detection

Telemetry is the auto recording and transmission of collected data from inaccessible or remote sources to an IT system in a another potion for monitoring and analysis, its gathers through sensors at the distant source which measures electrical (such as current or voltage) data or physical (such as pressure, precipitation or temperature), that provide the ability to monitor the state of an environment or object while it physically faraway.

Telemetry Tracking Command—Monitoring (TTC-M) is the way to monitor and control the satellite's functions and condition from the ground, as show in Fig. 1. Satellite telemetry data is a set of measurements taken on the satellite board and then transmit to the ground operations control centre, such measurements concern magnitudes such as voltages, currents, temperatures. Telemetry is a one direction communication from Satellite—to—ground, received continually during the entire mission life of the satellite, aims to monitor the satellite over situation reports and anomalies.

By analyzing the telemetry data will be able to make fault detection in the old data and fault prediction, an outlier analysis, anomaly detection, datasets segmentation, and datasets columns reduction.

Detecting the anomalies into the satellite telemetry data handled by multiple techniques, mainly split into two main categories: data-driven and knowledge-driven

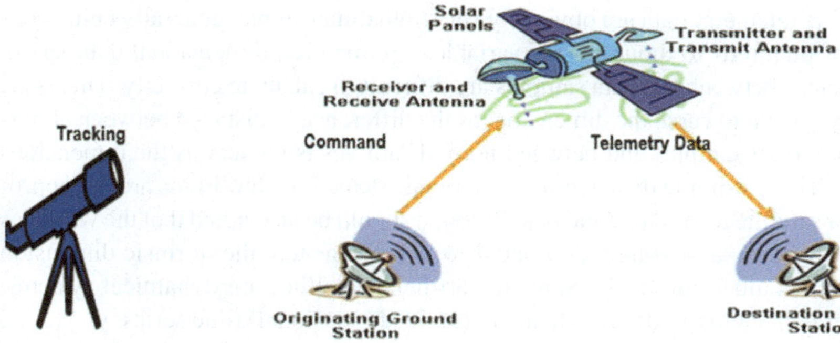


**Fig. 1** TTC-M satellite communications

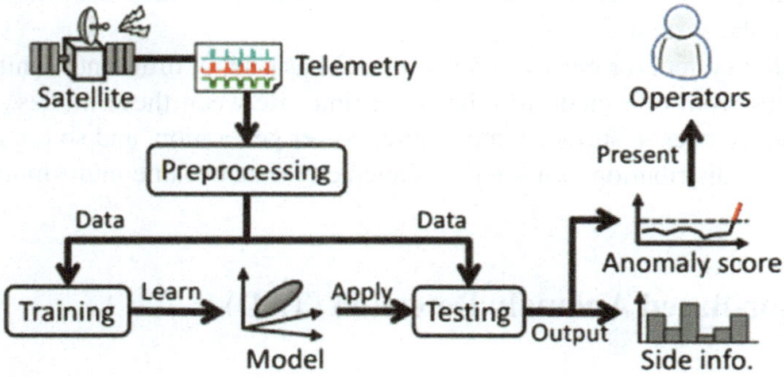


**Fig. 2** Data-driven anomaly detection flow [3]

approaches. The methods of the knowledge-driven approach built from experts' knowledge in advance, handles the anomalies detection and deduce the causes using: qualitative models, rule bases and probabilistic models. Such methods have the characteristic that they are capable to identify the anomalies in details, whether the knowledge is complete and accurate. However, it is very costly to set up and maintain such complete models and rule bases. For the data-driven approach, in the beginning learns the empirical models of the system by utilizing statistical machine learning algorithms into the old operation data, and after that, examine whether the system is normal or not through evaluating the most recent operation data employ the learned models. An assortment of machine learning methods, including regression, clustering, classification, kernel Statistical Principle Component Analysis (PCA) [1, 2], the hidden Markov model, and dimensionality reduction have been utilized for modeling space systems. Data-driven methods, most important feature is that they can be readily applied to a set of systems, science they do not need costly expert knowledge and auto learn the statistical models from the datasets. On the contrary, if a not sufficient volume of training data is presented, a data-driven method is not able to learn an appropriate model and is probable give miss true anomalies or many false alarms (Fig. 2).

Satellite telemetry data are obviously very high dimensional, generally being consisting of hundreds to thousands of variables. In this high-dimensional data space, the distances between the data samples are difficult to calculate correctly. This issue is widely known to curse the dimension, as the difference in distance between abnormal and normal samples and between normal samples is unclear as the dimensions increase. Thus, a simple distance-based anomaly detection algorithms are not appropriate for satellite telemetry data. In addition, it should be also noted that the variables in telemetry data are strongly correlated, which means that the intrinsic dimension of the data is much lower [4]. Since the artificial satellites are dynamical systems, the generated telemetry data by them are (multi-dimensional) time series. $y_t$, $y_{t+1}, \ldots$ There is no doubt that this time dependence is a fundamental feature of telemetry data and is very useful for system monitoring. For example, if the values of multiple variables change together at a time point, it is natural to consider that some event has occurred in the system.

A satellite system (or one of its subsystems) has a set of different running modes and changes from one mode to other over time. Between these modes, there are featured differences in surface temperature, power generation, and so on. As a consequence, the distribution of a satellite's telemetry data became multi-modal.

## 3 Tensor-Based Anomaly Detection (TAD)

A tensor is a geometric object utilized in physics and mathematics as extension of concepts like scalar, vector and matrix to higher dimensions. The root of the word 'tensor' is the Latin word *tendere* which mean 'to stretch'. Tensors could be represent as arrays of the format $X \in R^{L1xL2x \ldots xLN,}$ where N is the number of modes or orders. Therefore, tensors of order two, one and zero are, respectively, matrices, vectors and scalars. The analysis of tensors which has N > 2, i.e., three or multi-way arrays, is known as multi-way data analysis (Fig. 3).

TAD has been utilized within a diversity of disciplines over the recent years, since the research of MacGregor and Nomikos [5], researches associated to tensor-based anomaly detection (TAD) has been exponentially increasing. Moreover, much

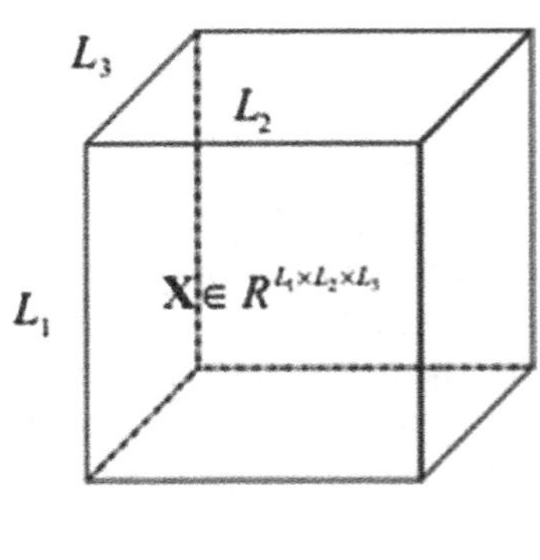


**Fig. 3** A third-order tensor

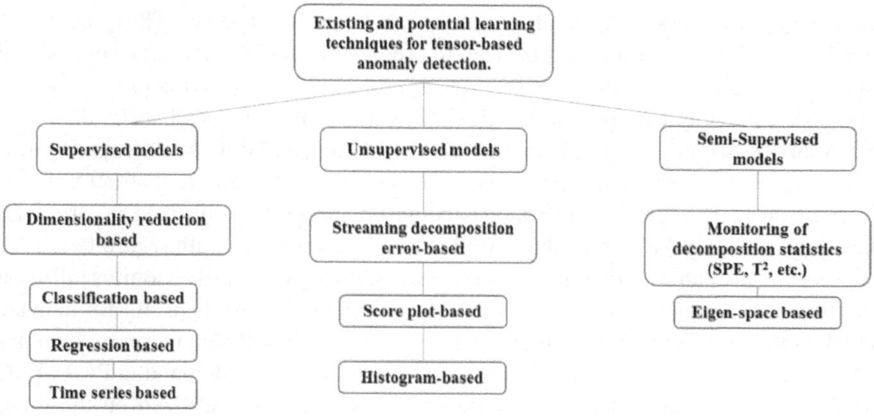


**Fig. 4** Tensor-based anomaly detection learning techniques

methods have been developed in multi disciplines from environmental monitoring and chemometrics to data mining and signal processing. Many existing and potential learning techniques related to the tensor-based anomaly detection are in detail into the interdisciplinary survey [6], summarized as the below chart (Fig. 4).

The following section, describe these models in more detail, with samples of the potential and exists work of these techniques.

## *3.1 Supervised Models*

Tensors has a significant role in dimensionality reduction for classification problems. For the time being, more learning methods are developed for supervised tensor-based. Several of these techniques, however of its potential for detect anomalies, are not yet utilized for this application.

### 3.1.1 Tensor Decomposition for Dimensionality Reduction

Tensors has an important role in dimensionality reduction for tensor decomposition, classification problems is employed as a dimensionality reduction tool for feature extraction, it consider a more advanced alternate for matrix-based dimensionality reduction solutions such as PCA. The proposed methods have two groups, fist one assumed that there are two sets, test and train (binary labels), where train set includes normal samples. Moreover, tensor decomposition is utilized on the normal tensor as a dimensionality reduction tool. Subsequently, one of the factor matrices (commonly time) is fed to an ordinary classifier (e.g. SVM or k-nearest neighbors) for building a model from the normal samples.

The target is to predict the observations labels into the test set. Thus, the developed model from the train set is utilized to predict the label (abnormal or normal) of observation into the test factor matrix. In [7] proposed to utilized the three-way data structure and apply a proper multi-way data analysis algorithm such as Parallel Factor Analysis, which is a simple model which obtained and utilized to train newness detectors. Such methods are evaluated both with simulated and real structural data to evaluate that the three-way analysis could be successfully utilized in structural health monitoring. Moreover, the advantage of such approach with regard to feature selection is also analyzed, Sensors make it possible to continually monitor pulses at multiple locations of a structure. Using a wide sensor network is useful for damage localization and a higher structural coverage, however it will also increase the number of variables. Thus, several dimensionality reduction is in demand, a PARAFAC decomposition accompanied by k number of components is utilized on the time-space- frequency tensor correlated to the normal samples and thereafter the acquired time factor matrix trained via k-NN (where features are the implicit variables). The model that has been built is then used for time point's classification in the incoming data.

Second methods group follows the selfsame procedure as the previous, however instead of bilateral labels (normal/abnormal) a numeric target is given for prediction. Thence, categorical classifiers replaces the regression models. Targets can be multiple or single variables. Bai et al. [8] develop an incoming supervised method in order to predicting earthquake ground movements into the domain of wavelet. The training input is a collection of seismological predictors related to path, local site conditions and seismic source, moreover the training output composed of the weights from a multiway analysis of ground monuments. They handle wavelet transforms of acceleration records like images then extract essential patterns from them utilizing tensor decomposition. Then the extracted patterns decomposition weights joint to seismological variables utilizing general regression neural network (GRNN) (Fig. 5).

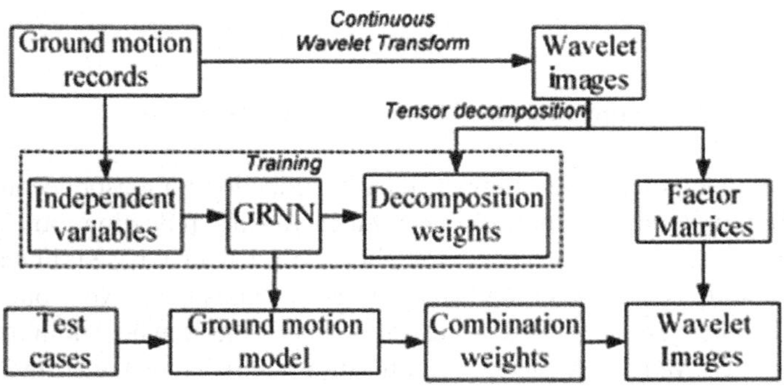


**Fig. 5** Diagram of the proposed procedure [8]

After that the produced nonparametric model will utilized to predict the accelerogram wavelet image for a provided set of seismological variables. Transform back the predicted image to the time domain is doable by using inverse wavelet transform for subsequent processing to correspond a given design spectrum. In contrast conventional ground motion models, moreover the proposed approach maintained the time domain features of ground motions. The utilized similarity metric between the vectorized forms of predicted and actual wavelet images in evaluating the prediction capability of the resulting model was Pearson's correlation coefficient. The results of the experimental evince the ability of the proposed model in order to predict important patterns into the seismic energy distribution. This type of approaches may be facilely extended for detecting anomalies. Another step, however, is wanted. For example, the difference of actual and predicted values can be utilized along with a threshold to anomalies detection.

### 3.1.2 Tensor Classifier

The regular classifiers adapt for tensorial data to be Tensor classifiers. Within these methods, datasets is trained directly through tensor-based classifier so then the developed model is utilized for prediction. A binary tensor classifier has a great capacity for detect anomalies from multiway data. Zhang et al. [9] Suggest a new method to represent an image object as a multifeature tensor that contain both the textural and spectral information (Gabor function), they presented a method wherever support vector machines (SVM) is expanded to support tensor machines (STM). The novel tensorial classifier is directly trained with the tensorial data of specified objects and then the developed model is utilized for target detection (Fig. 6).

Tao et al. [10] proposed a generic framework named Supervised Tensor Learning (STL) which adjust a lot of classic machine learning technique to occupy higher order tensors as inputs. The developed model tested successfully for the binary classification problems that can be very helpful for detecting anomalies. The Supervised tensor learning framework is a mix of the operations in multilinear algebra and

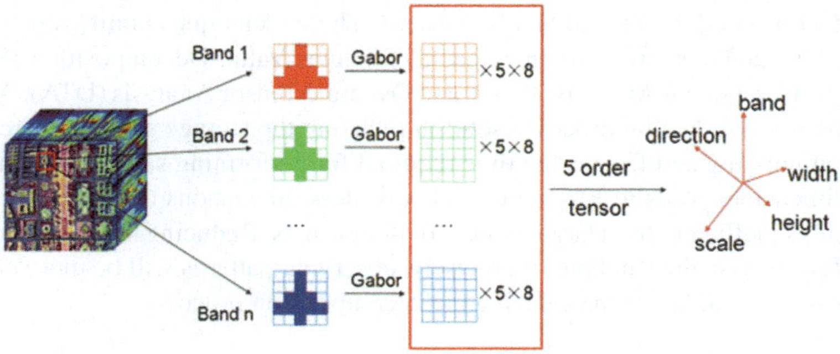


**Fig. 6** Representation of a remote-sensing image object as a five-order feature [9]

of the convex optimization. Tensor representation assist to decrease the overfitting problem into vector-based learning. Depends on the STL with its alternating projection optimization methods, they generalize the classic machine learning techniques like: support vector machines, Fisher discriminant analysis, minimax probability machine, and distance metric learning, to be support tensor machines, tensor Fisher discriminant analysis, tensor minimax probability machine, and the multiple distance metrics learning, respectively. To test the efficiency of STL, tensor minimax probability machine technique implemented by the authors for image classification, then comparing along with minimax probability machine technique, the tensor version reduces the overfitting problem.

The tensor-based learning and the vector-based learning have two different points between them: (1) the training measurements are represented via vectors into vector-based learning, while they are represented via tensors into tensor-based learning; in addition to (2) the classification decision function which defined by $\vec{w} \in \mathrm{R}^{\mathrm{L}}$ and $\mathrm{b} \in \mathrm{R}$ in vector-based learning $\left(\mathrm{y}(\vec{x}) = \mathrm{sign}[\vec{w}^{\mathrm{T}}\vec{x} + \mathrm{b}]\right)$, while the classification decision function is defined by $\vec{w}\mathrm{k} \in \mathrm{R}^{\mathrm{Lk}}$ $(1 \leq \mathrm{k} \leq \mathrm{M})$ and $\mathrm{b} \in \mathrm{R}$ in tensor-based learning, i.e., $\mathrm{y(X)} = \mathrm{sign}[\mathrm{X}\prod_{K=1}^{M} xk\ \vec{w}\mathrm{k} + \mathrm{b}]$. In vector-based learning, there are the classification hyperplane, i.e., $\vec{w}^{\mathrm{T}}\vec{x} + \mathrm{b} = 0$. While into tensor-based learning, the definition of the classification tensorplane, i.e., $\mathrm{X}\prod_{K=1}^{M} xk\ \vec{w}\mathrm{k} + \mathrm{b}$.

Cai et al. [11] Present a new method named Tensor Least Square (TLS) that is the expansion of least square classifier. Their experimental includes six databases from the UCI repository roved that tensor based classifiers are especially suitable for the small sample cases. This is because the fact that the parameters number estimated by a tensor classifier is more less than that estimated by a traditional vector classifier.

## 4 Tensor Decomposition

The process of converting a set of data having wide dimensions into data with minimal dimensions ensuring that it cover similar information briefly called Dimension Reduction. Tensor decomposition techniques which utilized into TAD could be categorized into main six categories of Tucker-based, Bayesian, PARAFAC-based, DEDICOM-based, LPPbased and ICA-based. These techniques family represented with multi methods, such as: Incremental Singular Value Decomposition (SVD), Principal Component Analysis (PCA), and Dynamic Tensor Analysis (DTA). Applying dimension reduction process useful in reducing the storage space required and data compressing and fastens the time required for performing same computations. Less dimensions leads to less computing, also less dimensions can allow usage of algorithms inefficient for a large number of dimensions. Reducing the dimensions of data allow to visualize and plot it precisely, observing patterns will be more clearly. Tensor rank could be estimate during the decomposition process.

## 5 Pervious Anomaly Detection Techniques for Satellite Telemetry Data

Dawei Pan et al. implement a data-driven anomaly detection technique for satellite sensor data, demonstrating integrated Kernel Principal Component Analysis (KPCA) with association rule mining. There are a total of seventy sensors deploying distributed into the satellite power subsystem. Their proposed method composed of three main actions:

- Extracting pattern from the multiple sensor data, thereafter mining association rules to each of the typical pattern existent in multiple time series.
- Analyzing the structure of measure space through its Eigen matrix by the KPCA together with temporal associated rules, furthermore discover the reason of anomaly by tracking the rules changes.
- Monitoring sensor real-time data from satellite power subsystem and detecting anomaly by KPCA method with associated rules.

They adopt the Piecewise Aggregate Approximation (PAA), as a linear segment representation technique in time series mining, to minimize data dimension. The below figure describe the framework of their proposed method (Fig. 7).

They utilize the sensor dataset MTS5000 $\times$ 63 from Feng-Yun satellite power subsystem. The training subset contains three thousand samples of sixty three parameters which are utilized to establish KPCA model. Further the testing subset contains the up-to-date data with anomalous samples produced by anomaly injection. The authors discover 113 associations through multiple sensor data from the satellite power subsystem.

L. Yuqing et al. propose a CUSUM control chart approach to handle the satellite power supply subsystem anomaly detection and fault early warning, by selection of the Power hydrogen pressure value as a feature from the satellites remote sensing data, then establish the CUSUM control chart from the power hydrogen pressure, thereafter detect the anomalies into the satellite power system using such CUSUM

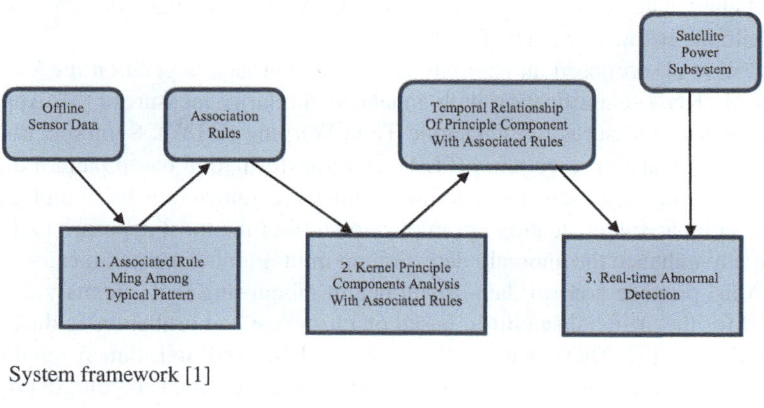


**Fig. 7** System framework [1]

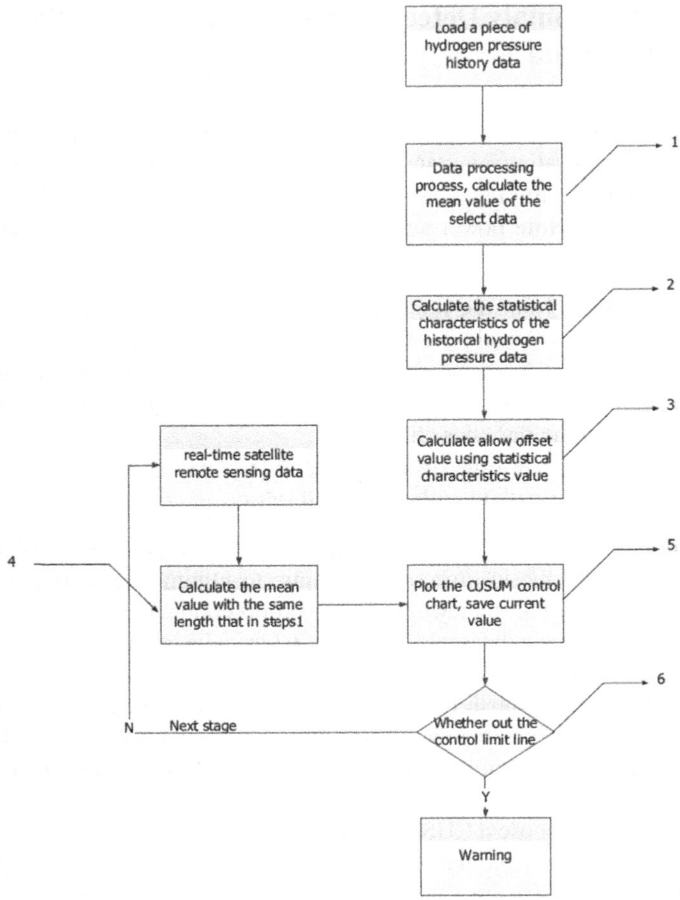


**Fig. 8** Method flowchart [12]

control chart. They describe the satellite power system's anomaly detection method steps into the below flowchart (Fig. 8).

D. Liu et al. proposed an anomaly detection approach depend on the k-Nearest Neighbor (KNN) classification with enhanced similarity measures [13]. Apply the new similarity measures like Dynamic Time Warping (DTW), Symbolic distance, Piecewise Linear Representation (PLR), and transformation based pattern distance in order to fully represent the satellite telemetry parameters. The comprehensive and the comparison evaluation are implicated to find the most appropriate distance measure to enhance the anomaly detection on multi monitoring parameters.

T. Yairi propose a novel data-driven health monitoring with anomaly detection method for the artificial satellites, based on clustering and probabilistic dimensionality reduction [3]. They focus on the multi-modality and high-dimensionality that are two significant features of the satellite housekeeping data, accordingly proposed a health monitoring-anomaly detection method depend on clustering and probabilistic dimensionality reduction in order to handle them. Secondly, they experimentally exercised the proposed method onto JAXA's Small Demonstration Satellite 4 (SDS-4) in operation, accordingly validated it over more than two years. The results of their experiment exhibit that the suggested data-driven monitoring method is very valuable, not just because it automatically detects the "anomalous" patterns which were hidden in the past, however also because it gives the operators with useful information in order to understanding the health status of the system with analyzing the causes of the discovered anomalies.

B. Nassar et al. present a proposed unsupervised learning algorithm based on Principle Component Analysis (PCA) technique for Space Telemetry Anomaly Detection [2], the algorithm introduces a functional approach for monitoring and diagnosis which includes: fault detection, fault diagnosis or identification and quality monitoring.

F. Bouleau et al. proposed an algorithm for an efficient outlier detection which builds an identity chart for the patterns utilizing the old data based on their fitting information curve [14]. His approach extract the features of the time series with enable traditional classification algorithms. Depending on the context, the data analysis may nevertheless differ and require re-classification. The proposed method provides fast data processing algorithms via utilizing synthesized information. Also has a methodology to compare two patterns utilizing the curve fitting information, along with the interesting properties. Furthermore measure the match quality, the used tools for horizontal identification and finally how the pattern's characteristics chart defined.

## 6 Tensor-Based Anomaly Detection Technique for Satellite Telemetry Data

Representing the satellite subsystems telemetry data as a multi-order tensor of measurements $\times$ space $\times$ time $\times$ mode, with the importance of tensors as novel category into spectral-based anomaly detection, enhance existing method for dimensions reductions or generate a new suitable one for the satellite telemetry data to be applied into supervised learning technique of the tensor-based anomaly detection.

Traditional popular methods for Satellite telemetry data anomaly detection are just capable to model second-dimensional data in addition they don't consider the cooperation between more than two dimensions. However, in the Satellite telemetry data, there is a shared connection between multiple dimensions, which lead to some meaningful anomalies may remain hidden with these methods.

An enhanced tensor-based technique below introduce as a novel approach for identification and detection of abnormalities and failures of the satellite telemetry data (Fig. 9).

The proposed framework include three main functions:

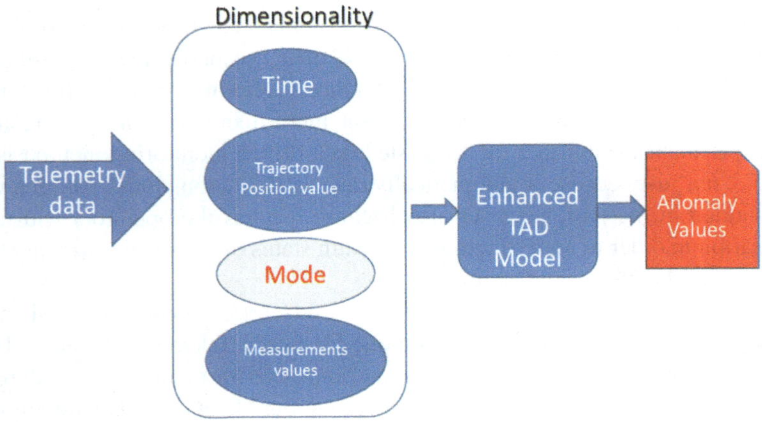


**Fig. 9** TAD for Satellite telemetry data

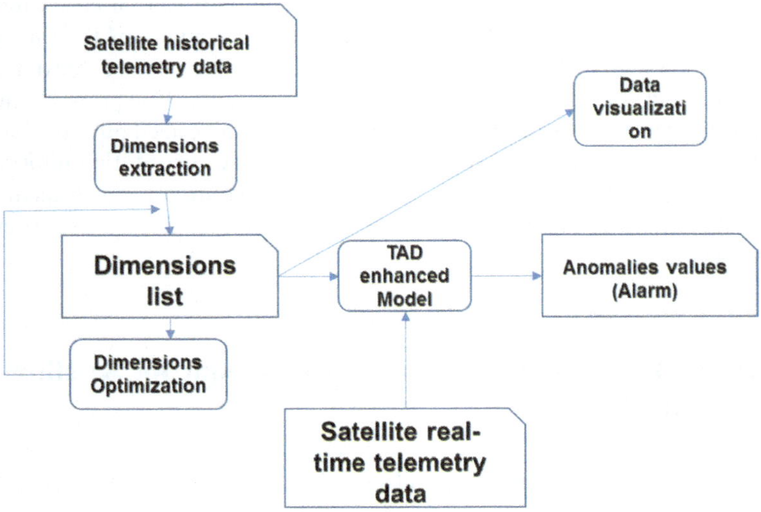


**Fig. 10** Proposed flowchart

- Data decomposition: extract all available related dimensions that represent the measurements of the subsystems historical telemetry data.
- Tensor optimization: use one of the swarm techniques to deduct the extracted dimensions, and keep only the minimum values that fully represent the data.
- Data visualization: visualize the tensor data allow the monitoring system and team to absorb the information quickly.
- Anomaly detection: find the anomalies values from the entered sub-system data, for failure protection and early warning (Fig. 10).

## 7 Conclusions

Detecting the anomalies values into a high dimensional-represented satellite telemetry data is the potential target, works on discover more meaningful invisible anomalies, which couldn't be detected by the traditional popular methods, in addition to overwrite some confront problems, such as overfitting, or large memory requirements.

Introducing the presenting of satellite telemetry data as a tensor data through the chain of decompositions and optimizations processes, to apply the enhanced TAD technique is the ambition goal of this work, to have an efficient warring and safety model for the Satellites. With study of the extended traditional machine learning techniques that support the multidimensional data in the supervised and non-supervised models.

**Acknowledgements** This research was supported by an TEDDSAT Project grant, Egypt.

## References

1. D. Pan, D. Liu, J. Zhou, G. Zhang. Anomaly detection for satellite power subsystem with associated rules based on kernel principal component analysis. Microelectron. Reliab. **55**(9–10), 2082–2086 (2015). ISSN 0026-2714
2. B. Nassar, W. Hussein, M. Mokhtar. Space telemetry anomaly detection based on statistical PCA algorithm (Version 10002768) (2015)
3. T. Yairi, N. Takeishi, T. Oda, Y. Nakajima, N. Nishimura, N. Takata, A data-driven health monitoring method for satellite housekeeping data based on probabilistic clustering and dimensionality reduction. IEEE Trans. Aerosp. Electron. Syst. **53**(3), 1384–1401 (2017)
4. A. Zimek, E. Schubert, H.-P. Kriegel, A survey on unsupervised outlier detection in high-dimensional numerical data. Stat. Anal. Data Min. **5**(5), 363–387 (2012)
5. P. Nomikos, J.F. MacGregor, Monitoring batch processes using multiway principal component analysis. AIChE J. **40**(8), 1361–1375 (1994)
6. H. Fanaee-T, J. Gama, Tensor-based anomaly detection: an interdisciplinary survey. Knowl.-Based Syst. **98**, 130–147 (2016). ISSN 0950-7051
7. M.A. Prada, J. Toivola, J. Kullaa, J. Hollmén, Three-way analysis of structural health monitoring data. Neurocomputing **80**, 119–128 (2012). https://doi.org/10.1016/j.neucom.2011.07.030
8. Y. Bai, J. Tezcan, Q. Cheng, J. Cheng. A multiway model for predicting earthquake ground motion, in *2013 14th ACIS International Conference on Software Engineering, Artificial Intelligence, Networking and Parallel/Distributed Computing* (2013). https://doi.org/10.1109/snpd.2013.17
9. L. Zhang, L. Zhang, D. Tao and X. Huang. A multifeature tensor for remote-sensing target recognition. IEEE Geosci. Remote Sens. Lett. **8**(2), 374–378 (2011). https://doi.org/10.1109/LGRS.2010.2077272
10. D. Tao, X. Li, X. Wu, W. Hu, S.J. Maybank, Supervised tensor learning. Knowl. Inf. Syst. **13**(1), 1–42 (2007). https://doi.org/10.1007/s10115-006-0050-6
11. D. Cai, X. He, J. Han, Learning with tensor representation. Technical report, Department of Computer Science, University of Illinois, (2006). UIUCDCSR-2006–2716

12. L. Yuqing, Y. Tianshe, C. Xueliang, W. Rixin, X. Minqiang. An anomaly detection algorithm of satellite power system based on CUSUM control chart, in *2016 3rd International Conference on Information Science and Control Engineering (ICISCE)* (Beijing, 2016), pp. 829–833
13. D. Liu, J. Pang, B. Xu, Z. Liu, J. Zhou, G. Zhang. Satellite telemetry data anomaly detection with hybrid similarity measures, in *2017 International Conference on Sensing, Diagnostics, Prognostics, and Control (SDPC)*, (Shanghai, 2017), pp. 591–596
14. F. Bouleau, C. Schommer. Owards the identification of outliers in satellite telemetry data by using fourier coefficients, in *Revised Selected Papers of the 6th International Conference on Agents and Artificial Intelligence - Volume 8946 (ICAART 2014)*, eds. by B. Duval, J. Van Den Herik, S. Loiseau, J. Filipe, vol. 8946 (Springer, Berlin, 2014), pp. 211–224

# Machine Learning in Satellites Monitoring and Risk Challenges

**Khaled Alielden**

**Abstract** The world we are living in is full of challenges such as climate changes, experiencing possible consequences of human-induced environmental change, increasing of population and depleting natural sources. All these challenges demand us to discover and understand comprehensively the natural sources to fulfill our needs. Evolution of sensors for collecting data with high resolution demanded us a new technology to handle the obtained big data to make a better decision. Machine Learning (ML) and Artificial Intelligent (AI) have a vital impact on the evolution of many sectors in economy and human services to develop our daily life. Satellites today service many sectors such as weather services, navigation, space-based telecommunications and direct broadcasting. Health monitoring of Satellites depend mainly on handling the large amount of delivered telemetry data using machine learning techniques for understanding the nature surrounding us to improve our systems and protect our environment. This chapter will show different aspects of satellites systems, the various orbits of satellites between and the risk challenges on satellites operation. Also, it will discuss the uses of satellites in image processing and the importance of machine learning to understand and work out the facing challenges.

## 1 Satellite Orbit

### *1.1 Different Orbits of Satellites*

Satellites use in many applications for assisting human trends in various research studies. The orbit location and the inclination angle of satellite is determined upon the satellite proposed application and the area, it is desired to serve. Each orbit distinguishes from the others according to its altitude which the latter demands a specific velocity of satellite in order to survive in its orbit. According to the satellite

K. Alielden (✉)
Physics Department, Helwan University, Cairo, Egypt
e-mail: khaled.alielden@science.helwan.edu.eg



A. E. Hassanien et al. (eds.), *Machine Learning and Data Mining in Aerospace Technology*, Studies in Computational Intelligence 836,
https://doi.org/10.1007/978-3-030-20212-5_2

orbit and velocity, its attitude and covered proposed area for different applications is various in each orbit.

For instance, many communication satellites are settled on a geostationary Earth orbit (GEO). The latter orbit be nearly above the equator at 35,780 ± 20 km above Earth's surface, and the satellite orbits once a day synchronized with Earth rotation in the same direction as Earth and thus seems like stationary above specific point on the Earth surface. Also, satellites systems such as those used for broadcasting television settled on GEO. While the other satellites in the Low Earth Orbit (LEO) used for satellite phones applications and be approximately at latitude between 160 and 1200 km. The satellite orbit located between the LEO, and lower the Medium Earth Orbit (MEO), which in range 1200–35,800, mainly used for navigation systems such as GLONASS and Navstar or Global Positioning system (GPS) as shown in Fig. 1.

The actual satellite attitude and velocity in each orbit is because of many factors, some of these are assigned to space environments and outer space and some are due to physical characteristics of the Earth, i.e. mass, density and so on. The satellite orbits the Earth affected by the force of Earth's gravitational field. This force pulls

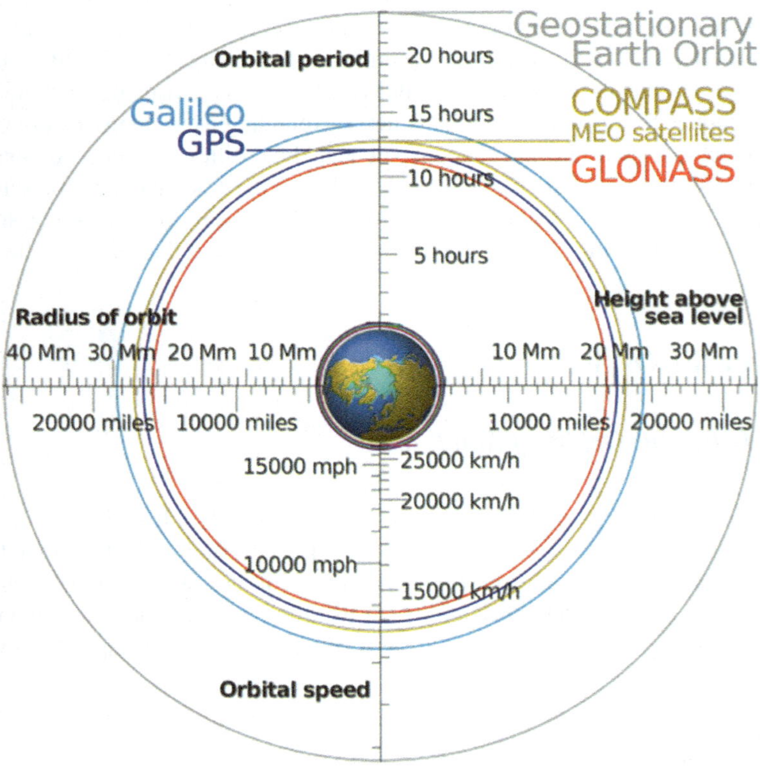


**Fig. 1** Shows the different orbits of satellites and their corresponding speed and period

the satellite back in toward the Earth. To overcome this force, the satellite, the satellite orbits with no lower than specific velocity called "scape velocity" ($v_{esc}$). Otherwise, the satellite will fall or movie in an orbital decay and burn up in the upper layers of atmosphere. Furthermore, the satellite is influenced by a pushing force due to its closely rotation around the Earth. This force so-called "centrifugal force" pushes the satellite away from the Earth. For any orbit, there is a given velocity which balance gravity force and the centrifugal force and remains the satellite in a stable orbit. This velocity is a calculated according to the formula

$$v_{esc} = \sqrt{\frac{GM}{r}} \tag{1}$$

where $G$ is the universal gravitational constant, $M$ is the mass of the body to be escaped from which in this case is Earth, and $r$ the distance from the center of mass of the Earth to the satellite. Equation 1 depends only on the altitude of the satellite orbit. Obviously, for orbit at a very low altitude, the gravitational pull is stronger, and this require the satellite to movie faster to counteract this pull. For instance, the satellite at orbit of around 100 miles must rotate with velocity 17,470 mile/h around the Earth that means the satellite complete one cycle around the Earth every 90 min as following

$$v_{esc} = \sqrt{\frac{6.674 \times 10^{-11} \times 5.97 \times 10^{24}}{(6371 + 160.9) \times 10^{3}}} = 17{,}470\,\text{mile/h} \tag{2}$$

$$t_{rot} = \frac{2\pi r}{v_{esc}} = \frac{2\pi \times 4058.17\,\text{mile}}{17{,}470} = 1.459\,\text{h} \cong 90\,\text{min} \tag{3}$$

For orbit at an altitude of 22,000 miles, the rotating velocity is about 6908 miles per hour which means the given orbit time of about 24 h. The satellite orbits in a circular orbit with the same speed whatever it rotates in the same direction as the rotation of the Earth (called "Posigrade") or it rotates in the opposite direction (called "Retrograde"). However, in the elliptical orbit, the speed of satellite changes upon in its position in the orbit. It reaches a maximum speed when it is at the closest point to the Earth and the minimum speed when it is at the farthest point from the Earth.

## 1.2 Different Uses of Satellites

Our ability to meet the challenges mostly depends on understanding the natural of Earth system and utilize that information to take the right decision. In this section, an overview of the remote-sensing, which means satellite remote sensing system will be discussed with a comprehensive view of the platform and sensor system. The system of transmission and receiving data including the processing and analysis of the acquired data for mapping category variables and generating high-level validation of product system, and remote sensing applications. The data acquisition system consists of the platform and the sensors on which the sensors reside, and the platform may be on the surface, or in space. Satellite sensor systems are orbiting in a geostationary or polar-orbit.

A geostationary satellite is in a geostationary orbit (GEO), which appears to the ground observers that the satellite fixed over one longitude and limited to approximately 60° of latitude at the equator; at a fixed point in the sky. In such case, the ground stations do not need to track the satellite and costs decrease. Also, due to the satellites continuously cover a large portion of the Earth in GEO, it is considered an ideal orbit for telecommunications and for monitoring continent-wide weather patterns and environmental changes. For instance, at least a constellation of three equally spaced satellites can provide full coverage of the Earth, except for the polar regions. There are several hundred communication satellites are used for coverage data like voice, data and video and several satellites which have very high-resolution radio meters payloads for weather forecasting and meteorological purposes. In addition, the satellites at Medium low earth orbit (MEO) orbits with speed approximately 16,330 miles per hour at an altitude of around 1000 km. this orbit is also particularly suited for telecommunications satellites.

Polar-orbiting satellites pass over the Earth's polar regions from north to south once every 100 min. The satellites in the polar orbit pass a track within 20–30° of the poles and do not cross the poles exactly. This is since near the poles, the probability of damaging the satellites increases as a consequence of the vulnerability to the injection of energetic particles from the outer space or from trapped particles in the earth magnetosphere into the earth at the polar regions. This will be discussed herein later. Satellites in the polar orbits pass over the North and South Poles several times a day and mainly take place at LEO altitudes i.e. between 200 and 1000 km. These satellites have a determinant swath width of path to cover the surface of the earth. In other words, these satellites orbits revisit same local time on specific location synchronous with sun illumination (known as sun synchronous orbit). That means the orbit shifted about 0.986° every day longitudinally to follow sun which means 360° in 365 days. That low orbits allow satellites in polar orbit to look down on the Earth's entire surface and collect data at high spatial resolution than from geostationary satellites. The altitude of sun synchronous orbit is between 600 and 800 km. Generally, polar orbits are usually used for weather forecasting, solar study, Earth observation like remote sensing and reconnaissance purposes. This to show a complete picture of the state-of-the-art development of remote sensing data processing techniques for different proposes by linking the chapters in the rest of the book and filling in any possible gaps.

# 2 Satellites Monitoring

## *2.1 Satellite Remote Sensing*

Over the past few decades, there is a revolution in the remote sensing science. It is considered a tremendous source of information needed by policy makers, resource managers, and vital for forecasters and sustainable future management of the Earth. A remote sensing system consists of sensors, processing, and analysis designed to monitor the variations, and forecast evolution of the physical and biological patterns of the Earth system. There sensors are categorized to passive and active. Passive sensors detect the radiation that is emitted by the object or reflected from the object from an external source other than the sensor like sunlight. Typical passive sensors include radiometer, imaging radiometer and spectroradiometer.

(1) Radiometer is an instrument that measure the radiance of electromagnetic radiation in a spectrum region from microwave to visible light.
(2) An imaging radiometer is a radiometer scanner for providing a 2D array of pixels to gain an image. This scanning or imaging process can be performed electronically or mechanically by using an array of detectors which classified into alone-track or across-track scanners. The lone-track scanner or Push broom scanner, consists of a linear array of Charge Coupled Devices (CCD) arranged perpendicular to the flight direction of the spacecraft without using a mechanical rotation device as used in a whisk broom scanner (will be discussed next) [1]. The latter imaging different areas of the surface or to be more obvious, it scans a swath as the spacecraft flies forward. A push broom scanner gathers lighter than a whisk broom scanner because it focuses on a particular area for a longer time. This kind of scanner has a low resolution due to the varying sensitivity of the individual detectors. Across-track scanner or Whisk broom Scanner, scanning from one side of the sensor to the other across the spacecraft flight direction using a rotating mirror. The mirror scans and reflects light into a single detector which collects data one pixel at a time. Whisk broom scanners can focus the detectors on a subsection of the swath width by stopping the scan. This advantage is typically for a high-resolution imaging comparing with a push broom imaging that designed for scanning the same size of swath. The drawback of this type of scanner sensor is its expensive costs.
(3) Spectroradiometer: A radiometer that measure the radiance or irradiance in multiple spectral bands, such as the Moderate Resolution Imaging Spectroradiometer (MODIS) and the Multi-angle Imaging Spectro Radiometer (MISR). These data provide a necessary information for understanding the dynamics and processes that going on the land and in the oceans, and in the lower Earth's atmosphere.

Active sensors send a pulse of electromagnetic radiation to the object they observe to illuminate it and then receive the reflected radiation from that object. Typical active sensors include radar, Synthetic-aperture radar (SAR), Interferometric synthetic aperture radar, Scatterometer, Lidar and Laser Altimeter.

(1) Radar (Radio Detection and Ranging) transmit pluses of electromagnetic radiation in microwave range and record the time of the reflected or backscattered pulsed radiation from distant objects to calculate the distance of the object.
(2) Synthetic-aperture radar (SAR) spaceborne is a side-looking radar, incidence or angle θ, imaging system that images Earth's surface as shown in Fig. 2. The SAR transmits a series beams of electromagnetic microwave pulses during its motion, which is assumed straight path in a small length scale for simplicity as shown in Fig. 2, that illuminate a swath width on the ground. Then the SAR receiver detects and synthetic the echoes reflected signal from the ground for producing high spatial resolution imagery. The backscatter echoes, which scatter back toward the SAR receiver, are produced due to interact of microwave radiation with the different terrains on the ground such as buildings, trees, mountains and rocks.
(3) Interferometric synthetic aperture radar (InSAR) technique compares phase difference between two SAR images to generate a map of surface for the same geographic region at different times which received during travel of SAR. The advantage of using the phase difference between two SAR images in InSAR is height information/resolution to detect millimeter to centimeter scale ground deformation resolution and 30-m pixel resolution and covering areas about 100 km × 100 km [1]. It uses for geophysical monitoring temporal ground surface changes and natural hazards, such as earthquakes, volcanoes and landslides (Fig. 3).
(4) Scatterometer is an instrument designed to detect the backscattered microwave signal power and determine the normalized radar cross section of a surface. Scatterometer have been used to measure and derive maps of surface wind

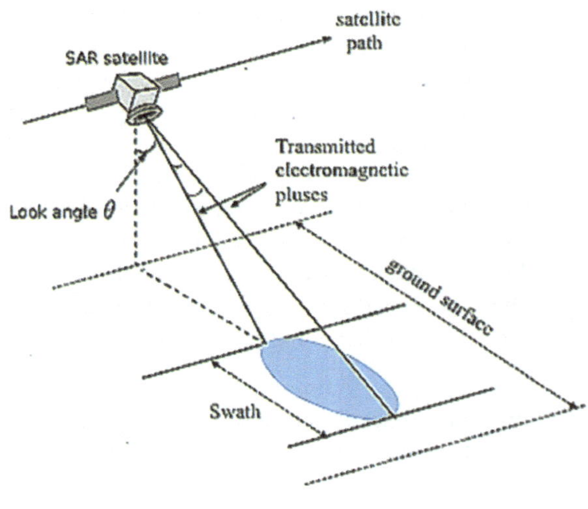


**Fig. 2** Schematic geometry of SAR system

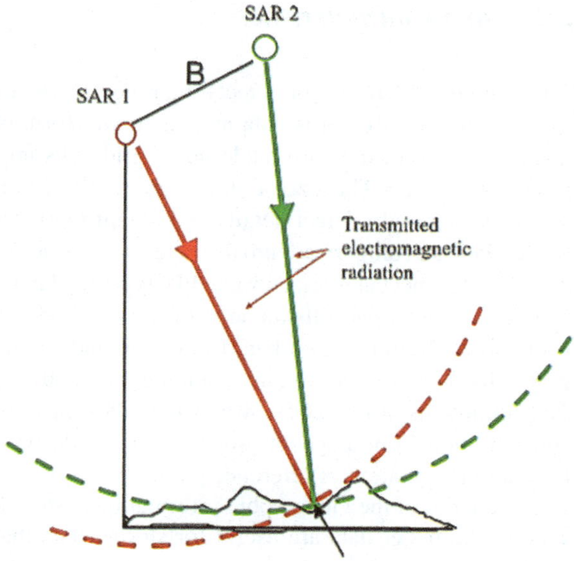


**Fig. 3** Shows the Interferometric technique. Return comes from intersection

speed and direction over sand, oceans and snow dunes from space. It has also been used for mapping surface soil moisture and freeze.

(5) Lidar (Light/or laser Detection and Ranging) is a remote measurement technique that uses sensitive optical sensors to detect and analyze the backscattered beam of light. Lidar transmits higher frequency pulses such as ultraviolet, visible, or near infrared spectrum rather than radio waves that uses in radar. One of the applications of Lidar is determining the distance of the object by knowing the speed of light and recording the delay between the transmitted and backscattered pulses. Also, by knowing frequency of the transmitted and the backscattered pulses, it is possible to determine the speed of distant object using Doppler effect. Furthermore, if it is possible to isolate the interaction of light or laser with the matter in a diffuse media such as the atmosphere, the physical parameters of the gas such as density, temperature, etc. could be estimated and specific gas could be identified.

(6) Geoscience Laser Altimeter System (GLAS) is an instrument combines laser altimeter for continuous global observations of Earth. The Laser Altimeter system is a group of active remote sensing techniques that use a Lidar for measuring the height of instrument platform with respect to the Earth's surface and determine the topography of the distance surface.

## 2.2 Data Characteristics

The remote sensing sensor detects the reflected electromagnetic radiation from the Earth's surface. It records data as numbers in form of raster image data as shown in Fig. 4a. Raster data are made up of grid cells are called pixels and each pixel has its own value. The size of an area represented in a pixel determine the capability of the sensor to detect details and obtain more resolution as shown in Fig. 4b. Raster data are categorized into discrete or continuous. In the discrete raster, the grid cell has a distinct categorized value like type of a land cover and type of soil. In other words, the data value fills the area of the pixel and each data type, like and cover, is classified discretely to urban, forest, soil and so on. In the Continuous raster, the grid cells data have a gradual changing value such as elevation, temperature etc. The continuous raster can show how the fluids move from high concentration to low concentration from a specific source and can derive an elevation model using sea level as a registration or reference point.

In general, in the Geographic information system (GIS), data types are roster or vector. The roster data are useful for storing data that varies continuously such as satellite image and remote sensing data, surface of chemical concentrations, or an elevation surface. However, Vector Data represents the surface as points, lines, and polygons as shown in Fig. 5. It is useful for storing data that has discrete boundaries, such as country borders, land parcels, and streets.

The specifications of the sensors that used in remote sensing applications reflect the data resolution like spatial, spectral, temporal, and radiometric resolution.

(1) Spatial Resolution is the ability of a sensor to detect details. In other words, it is the sensor ability to resolve the smallest distant object, or the ground area imaged for the instantaneous field of view (IFOV) of the sensor. The Spatial resolution of images are frequently expressed in meters. For example, the satellite sensor acquired image with "10-m" resolution means that two objects can be separated

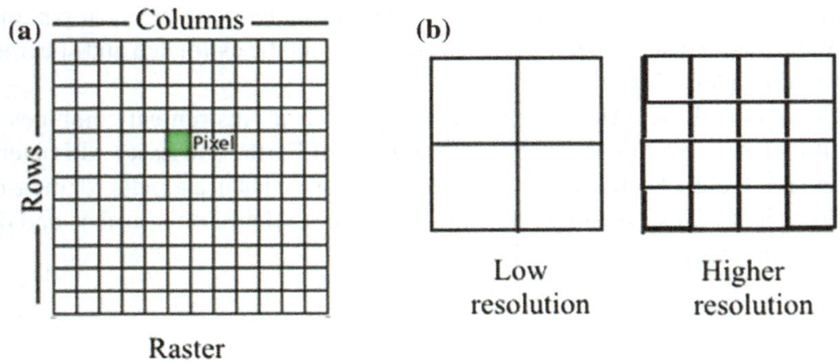


**Fig. 4** Shows the raster data as recorded by sensors. **a** Shows the grid cell for restoring the raster data. **b** Shows the different between low resolution and high resolution of the raster data

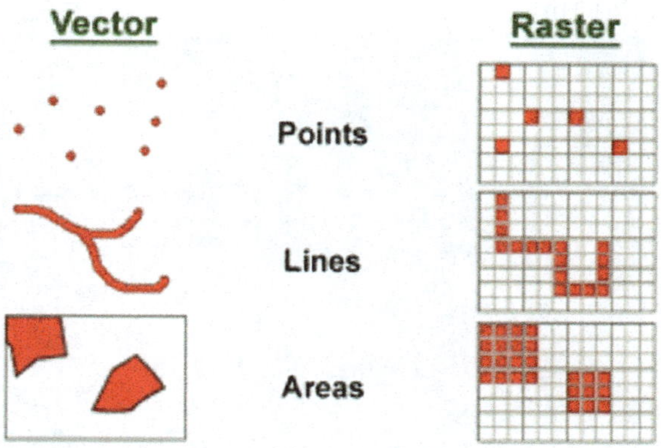


**Fig. 5** Shows the difference between the raster and vector data

if they are sitting side by side can be, ten meters long or wide. Less than ten meters, one couldn't separate between them as shown in Fig. 6 [1].

(2) Spectral Resolution is the ability of a sensor to distinguish the differences in wavelength or width of spectral bands in a sensor system (see Fig. 7). Many sensor systems have a panchromatic film, which is sensitive to a wide range of wavelengths (one single wide band in the visible spectrum, and multispectral bands in the RGB (abbreviation of red, green, blue), NIR, Mid-IR, or thermal-IR spectrum). Combination of spectral bands is useful for identifying features of ground surface (see Fig. 8). The system that have hundreds of spectral narrow bands called hyperspectral systems.

(3) Temporal Resolution is a measure how often the sensor repeats coverage or cycle to revisit the same part of the Earth's surface and repeat measurement. The Temporal Resolution characteristics depend on satellite orbit and sensor platform. The temporal resolution is used for temporal consideration during day, year or season such as seasonal differences, tidal stage, leaf on/leaf off etc. For instance, Fig. 9 shows the seasonal temporal consideration between spring and summer.

(4) Radiometric resolution is the ability of the sensor to distinguish between the magnitude of the electromagnetic energy. In other words, the ability of the sensor to discriminate small differences in the magnitude of radiation within the ground area for a single raster cell. The higher radiometric resolution of the images is determined by the higher number of data bits that sensor records per pixel. Obviously, imagery data are represented by positive digital numbers which vary from 0 to a selected power of 2. This range corresponds to the number of bits used for coding numbers in binary format. Each bit records an exponent of power 2. Therefore, the maximum number of brightness levels, that represent the energy recorded, mainly depends on the number of bits per pixel. Thus, if a sensor used 10 bits to record the data, there would be $2^{10} = 1024$ digital values available, ranging from 0 to 1023 for each pixel (see Fig. 10) [1].

(a) 1m

(b) 10m

**Fig. 6** Shows images of the same area as detected by difference spatial resolution **a** at 1-m resolution. **b** At 10 m resolution

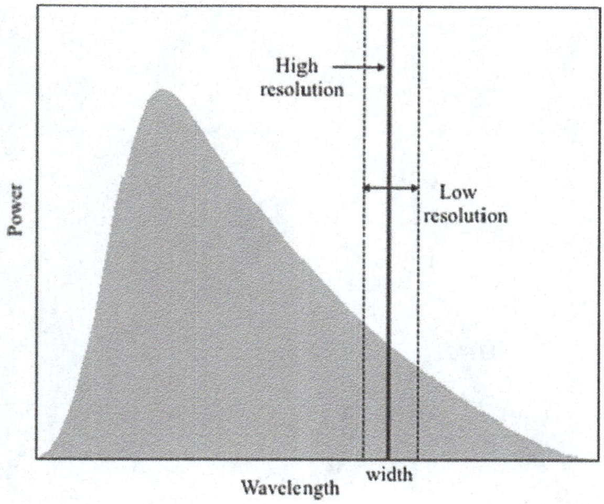


**Fig. 7** Shows that the low-resolution sensors record energy within relatively wide wavelength bands (represented by to dashed lines). However, the high-resolution sensors record energy within narrow bands (represented by solid line)

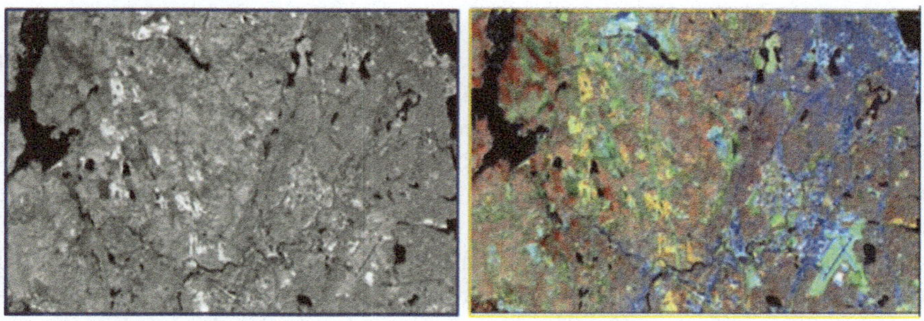

**Fig. 8** Show the same are as observed by sensor. The left one shows the Landsat-7 Panchromatic Data (15 m) and the right one shows the Landsat-7 ETM+ Data (30 m), Bands 4, 5, 3 in RGB

Satellite remote sensing technology and the science associated with evaluation of its imagery data provides potentially valuable information for assisting human research studies in various dimensions. In generally, remote sensing is the science of identification and estimation of the physical properties of distant objects using reflected, scattered or emitted electromagnetic radiation. The spatial, temporal and polarization spectral from the object are all signatures and discriminate the characteristics of the object. All the Space and Astronomy researches depends on the remote sensing science. Herein, we focus on the satellite remote sensing on Earth. The imagery data from satellite remote sensing that taken in different wavelengths are processed before extraction of spectral information. Observe a synoptic view at different resolutions and repetitive coverage with calibrated sensors to observe changes, provides a better chance for natural resources management. The Satellite

**Fig. 9** Shows the temporal consideration for the same image in different season. The left-hand side image shows the area in spring in band 4, 5, 3 RGB. The right-hand side image shows the area in summer in band 4, 5, 3 RGB

remote sensing has provided imagery data of environmental nowcasting changes at inaccessible locations in oceanography, agriculture, geology, meteorology and disaster control, etc. The information gathered from satellite remote sensing are utilized to predict future patterns and help in making better decisions to utilize the environmental and achieve the best outcomes in different areas which have effects on the economics and political decisions.

For Instance, in meteorology, the intensity of Earth's solar radiation, geothermal energy and dynamic of winds are monitored and measured by satellite remote sensing for acquiring weather information, exploring and managing the energy resources. Satellite imageries are also used for improving models for forecasting natural disasters such as air disasters, floods and earthquakes and estimating damages, including catastrophic events and provide appropriate warning. In agriculture, it is used for identifying potential threats on crops and understanding well the water cycle that leads to improving water resources management which necessary for life and crop growth. Satellite remotely images have allowed global mapping and monitoring changes in oceanography such as topography of surface area, phytoplankton content, currents and winds, which useful for establishing habitat linkage between oceanographic processes and fishery resources. Moreover, it utilized in glaciology, where it allows monitoring the temporal dynamics of glaciers. In geology, it used for exploring and identifying the composition of minerals in the ground. The applications of satellite remote sensing are increasing, due to its speed in gathering the information and efficiency as well. It becomes a necessary tool in environmental resource management process for knowing the effects of environmental factors on human health and well-being.

(a) 8 bits (256 digital values)

(b) 4 bits (16 digital values)

**Fig. 10** Shows two radiometric resolutions for the same area

# 3 Risk Challenges

## *3.1 Space Weather Impacts*

A mere handful of satellites could not withstand so-called "space weather impacts" which mainly related to solar activity. These impacts including digital systems of satellites, satellite navigation systems, radio technologies, and major effects on the operation of satellites in low Earth orbit that known as drag effects. The four major solar events that affect on satellite communications components are coronal holes, solar wind, coronal mass ejections (CMEs) and solar flares. The maximum and minimum of solar activity is predictable according to solar cycle which repeats every 11-years. The solar activities beside the disturbances in Earth's magnetosphere and all different sources that affect or occur near Earth is called space weather effects.

The solar wind is approximately constant but varies in velocities and intensity for a while, whilst the other three solar phenomena come and go. The abruptly changes, without warning, in the performance of satellite components is called single-event upsets (SEUs). It causes by high-energy protons and heavier ions (>10 meV) which generated by CMEs, solar flares or accelerated by shock waves. The SEUs are not apt to be caused by the solar wind and seldom penetrates the outer protective layers of a spacecraft because of its relatively low in energy comparing to the energies of solar energetic particles which generated from/or associated with coronal holes, solar flares and CMEs which their powerful disruptive the satellite's components and extends to long distances in the heliosphere. The CMEs may not impact satellites at all, due to the curving trajectory of CMEs influencing by the interplanetary magnetic field lines but, its magnetic field can affect on satellites power systems or generate shock waves that accelerate the particles and affect harmfully on the satellites components. Solar flares are huge intensities of energetic particles and quite flash of X-ray explosions on the sun. They eject a lot of charged particles mainly protons. All these events increase the levels of energetic particle radiation in space near the Earth's atmosphere. These particles disrupt the digital systems of satellites either by penetrating directly into the satellite electronics, or through charging of spacecraft that generate discharge issues which damage electronics and even loss of control on the satellite. For instance, when high energic particles/ions plough through electronic chip or digital devices of the satellite component, a single high-energy particle can deposit electrical charge in sensitive regions of the device like memory cells. This effect so-called "single-event effects (SEE)" increase the numbers of electron-hole pairs that carry currents within these devices. This process known as the single-event upset. As the deposited charge is sufficient to alter a bit from 0 to 1 and or vise verse which, consequently, alter the data stored in the device. Also, this effect degrades semiconductor lifetimes.

Spacecrafts in synchronous orbit expose to charging process which includes both surface charging and internal dielectric charging. The spacecraft may negatively charge because of the abundant of electrons in the inner magnetosphere during geomagnetic substorms. Surface charging occurs due to incidence of a large incoming flux of low-energy plasma and geomagnetic substorms which create photoelectric currents in the absence of effective mechanisms of charge drainage. Surface charging anomalies occur more often in the dark side of the satellite orbit i.e. midnight to dawn sector. Obviously, in the absence of sunlight i.e. during eclipse or due to the injection of low-energy electrons (<100 keV) into the Earth magnetosphere. However, in the day side of the satellite orbit, the satellite emerges into sunlight and a potential discharge is created i.e. a positive surface charge is formed due to photoelectron emission.

The other kind of charging is Internal Dielectric Charging which caused by relatively high energetic electrons (>100 keV) penetrating dielectric materials. Most of these relatively high energetic electrons are trapped by earth's magnetic field and from Van Allen radiation belts precisely. The occurrence of this create internal charging of satellite components by energetic electrons and increase the electron density. Most of high charging induced anomalies are assigned to deep dielectric charging than from the surface charging or SEUs. Electrostatic discharge (ESD) occurs due to charging. Once the electric field exceeds a threshold $10^{10}$ electrons/cm$^2$, an arc discharge will be build-up and generate an electromagnetic transient that may interact with satellite electronics and causes satellite operational anomalies or even complete failure of the satellite. Internal discharge is devastating since it occurs within dielectric materials and the generated arc as a pulse of widths in tens of nanoseconds and appears on the cabling and circuit board. In Other words, if charge buildup occurs more rapidly, then the probability of the arc discharge occurring increase. In generally, most of the undesired discharge effects that cause satellite operational anomalies are the rising of discharge arc, damage of the physical materials and generation of electromagnetic interference (EMI). Many manners are used to restrict the satellite charging like shielding to reduce the probability of penetration and causes of internal charging. In addition, EMI-susceptibility reduction techniques can be employed to mitigate the effects of arcing. Furthermore, strong magnetic field storms in the geostationary orbit can cause disruption in the satellite systems. On Earth, Space weather effects on ground can disrupt power distribution networks, railways and increased pipeline corrosion and cause degradation of radio communications and drag of satellite as discussed next.

### 3.1.1 Radio Systems

Furthermore, the space weather impacts on the radio communication systems with satellites. For instance, energetic particles and X-rays that generated from solar flares, can increase the density of electrons and ions in the ionospheric layer at low altitudes around 70–90 km. This denser layer has higher critical frequencies that can reflect low-frequency radio signals back down to the ground and they can interfere with the signals propagating as an interface wave. For the Global Navigation Satellite Systems (GNSS), this impact on the propagation of satellites signals down toward the ground cause delay in the arrival time of the signals which leading to errors in determining the position.

### 3.1.2 Satellite Drag

Space weather has dramatic effects on the deviation of satellite from its orbit. The space weather impacts on the uncertainty in the awareness of satellite's location and existance of debris in low Earth orbits (LEO). This topic of research called space situational awareness. Large changes occur during the space weather activity. For instance, The EUV, X-rays that preceded the violet explosion on the sun can penetrate the Earth's atmosphere and ionize its particles. The ionized particles and particle precipitating in the Earth's magnetosphere change the thermospheric density over polar regions. In addition, during the geomagnetic storms or substorms, the aroura causes heating in the lower thermosphere over the polar regions. This thermal heating derives both an upwelling of denser material into the higher thermosphere and strong winds toward the equator. This denser material in the global thermosphere resist the travel of satellite in its orbit. This resistance exist on the satellite like a drag force acts opposite to direction of satellite motion and slow its velocity. This drag force pull the satellite closer to the earth and reenter the it's atmosphere (see Fig. 11). Furthermore, the thermal disturbance in the auroral region can create large-scale atmospheric gravity waves propagate toward the equator. These waves can create a measurable variation in the drag of a satellite during its travel through the peaks of dense matter of these waves [2]. The common examples of spacecraft operating in LEO are the International Space Station (ISS) and the Hubble Space Telescope.

To overcome this issue and survive the satellite in its orbit and don't loss the functionality it provides, the operators of satellites in these orbits forecast the location of their satellites, to plan operations for observing demand regions on the Earth's surface and schedule ground station contacts for uploading future operations adjust the satellite position and like obtain data from the satellite.

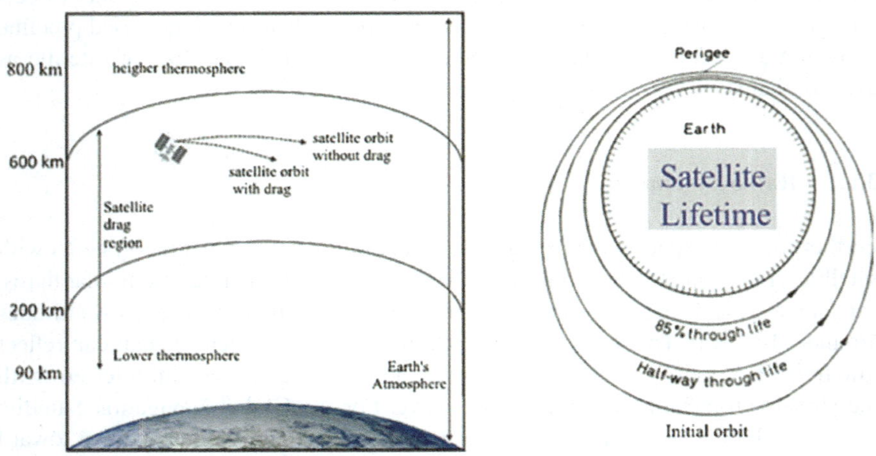


**Fig. 11** Shows the deviation of satellite from its orbit due to the drag of satellite

## 3.2 *Debris*

Space debris defined as every object that is non-functional including in our space environment. The most debris objects are human-made which generated, since human space age, from object in-orbit break-ups, a lot of explosions exceed 200 explosion, and few in-orbit collisions between objects. All these events produce fragments and elements that are orbiting and re-entering the Earth's atmosphere but no longer functional. Obviously, any collisions involving a working satellite can cause damages. These damages produce more space debris and the gradual increase of debris each decade increasing the collision risks for other satellites. The debris material could be millions of pieces ranging in size from 10 cm to smaller than 1 cm. Most of the debris is in low Earth orbit above the polar region, however, some debris can be found in geostationary orbit above the Equator. Thus, ISS work with satellite operators to execute a collision avoidance maneuver to avoid collisions that could create more debris. Figure 12 shows the evolution number of objects in Earth orbit [3].

How long a piece of space debris takes to fall back to Earth depends on its altitude. The debris at lower altitude than 600 km fall back to Earth within several years before re-entering Earth's atmosphere while debris at higher altitude than 1000 km orbit for centuries.

Different satellites are exposed to varying levels of risk according its location and orbit. In other words, the exposure of satellite to space weather impacts may vary depending on its orbit. The sought to overcome the challenges are finding new design systems with appropriate engineering solutions to diminish the risks

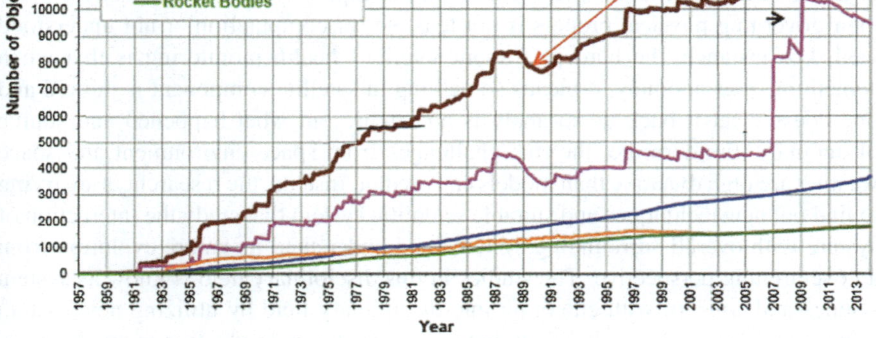


**Fig. 12** Shows monthly number of objects in Earth orbit by object type

posed by space weather or debris. Researches are going closely with industries to identify and understand the threat and doing forecasts that help them mitigate the damages. The researches in latter mentioned field is divide into two ways. The first one concerns with finding out the appropriate materials that may be exposed to much radiation without big damages and meanwhile decreases the possibility of deep penetration. These materials are used for shielding the satellites to survive and operate its components. In addition, find a suitable engineering designs and mechanisms that diminish the threats of the charging phenomenon. The other way is to monitor the anomalies of satellites systems and predict the malfunction that will occur. This work is known as health monitoring of the satellite and this work is crucial to identify the risk and predict the damages and thus take the right decision to protect the satellites.

## 4 Importance of Machine Learning and Applications

Machine learning (ML) is a tool of system science field that can learn from gathered sample input data and extract structural information for building a model. Expression of machine learning named by Arthur Samuel. The ML is mainly depends on computational statistics in which focus on inference of statistical models and make data analytics for optimizing the statistical models from data. In general, the learning of machine based on three principles: representation, evaluation, and optimization. Representation is a process of using the classifier elements in a formal algorithm that computer can handle and interpret the inputs. Evolution is a process of using functions to evaluate whether the classifiers are good or bad. Optimization is a process of searching, by formed algorithm, among the classifiers to find the highest score accuracy and performance.

The principle objective of ML is to build intelligent systems or machines that can pick up the structural information of the samples in the training sets and identify the variables that control the dynamic of the system from a comprehensive view. Also, to understand the behavior and modeling the complex systems especially for those their governing physical process is not feasible for computation or not understood well. For instance, the hard part about satellites health monitoring is that when something goes anomaly, it means something in satellite component is injured and one doesn't get it back to do analysis and figure out what happened and source of anomaly. Furthermore, the risk challenges from space environment and space weather are high dynamic than models can predict. Instead, the researchers are trying to find out new techniques and suitable scientific system that study the interaction of system with overall surrounding systems and understand its nature evolution from an overarching perspective. The studies in this disciplinary field is know as system science and concern with emerging interdisciplinary field by utilizing new system science techniques that rather than that used in traditional reduction methods.

The techniques that are followed in the system science is to reduce the complex system into approach components, then trying to understand each component and collaborate with other experienced components and thus develop a complete view of the whole system. In other words, the whole system is considered as a function unit of complex interacting systems. For instance, the comprehensive model that was designed by coupling the approach models of solar wind, global magnetosphere, and inner magnetosphere (e.g., SWMF; [4]). The system science handling with simulated or gathered measured data from real life for characterizing, building model that describes the behavior of the complex systems using this approach technique and finally forecasting its response [5]. Characterization is a task that identify patterns, dependency and degree of nonlinearity between variables of the system. While, modelling is a task that describe the evolution of the system by determining a suitable set of equations in which governing the dynamic of the system. Forecasting involves predicting the response of the model to identify variables which are responsible for state transitions. For Example, the model can predict the amount of damage that will occur after observing any anomaly in the satellite's systems attitude or the predict probability of debris collision occurrence to protect the satellite or predict the arrival time of hazard events the affect on Earth's infrastructure systems. Also, in remote sensing application such models use for fire detection, flood prediction and urban monitoring [6].

Machine learning and data mining are sometimes conflated for driving scientific modeling and are used in wide dimensions of natural science, engineering, medical science, social science and humanities. One of the most central questions when using machine learning is: what is the smallest central set of variables that is essential to describe the system. The set of these variables is known as state vector and the length of this state vector is known as the dimension of the system. Obviously, the state of variables is identified by observing the characteristics of the system. Settle on the variables are based on experience using technology, however, they are limited by understanding as well as temporal and spatial resolution. To build the state variable or the model that describes the system, there are multiple learning strategies that can be applied in machine learning process. These strategies can be categorized as: supervised learning, active learning, semi-supervised learning, unsupervised learning and reinforcement learning.

(1) Supervised learning is a technique that the algorithm learns with example inputs and their labeled data that were inputted by a human during the training process and the goal is to explore a general potential structural that drives inputs to desired outputs.
(2) Active learning is a technique that algorithm can interactively query the information source or user for labels. This technique is considered a supervised learning. In this case, the user chooses the training labels for a limited set of instances and the algorithm has to optimize its choice of objects to acquire labels. The learning takes place in a lower number of instances that required in normal supervised learning.
(3) Semi-supervised learning is a technique that algorithm learns from given only an incomplete labeled data during training process. In this case the algorithm

is trained on unlabeled data to define boundaries of those were not specified in the inputted labels by human.

(4) Unsupervised learning is a technique that algorithm learns from unlabeled data and the let algorithm looks at inherent similarities and discover hidden patterns in its input data to distinguish them into groups.

(5) Reinforcement learning is a technique that algorithm learns from reaction of the system in a dynamic environment such as driving a vehicle, control theory, game theory or swarm intelligence.

The different tasks of machine learning are categorized according to different propose for using it to obtain the desired output. The machine learning tasks used Classification, Clustering, Regression, Density estimation, Dimensionality reduction.

(1) Classification is a supervised learning task to extract a model that classify the inputs data into one or more classes, or label in case of multi-label classification. In this case the output data is discrete. For instance, task for classifying if the component will disrupt or not.

(2) Regression is a supervised learning task to extract model that describe the behavior of the system and use this model for predictions. In this case the outputs are continuous.

(3) Clustering is unsupervised learning task to divide a set of inputs into groups. It is an unsupervised task as the groups are not known beforehand, so it is not like the classification process. The clustering algorithm can be used to cluster analysis and gain deep insights from our data.

(4) Density estimation uses statistical models to find the probability distribution of inputs in some space like Kernel Density Estimation.

(5) Dimensionality reduction simplifies set of variables that captures the essential variations of the observed variables by representing them into a lower-dimensional space. This done by replacing by a subset of the observed variables or set new variables that better capture the underlying variation of the observed variables such as Principal component analysis.

Generally, the method is based on computational statistics and information theory. Let us consider two vectors of continuous random variables of output $\mathbf{Y} = Y_j$ for $j = 1, M$, regarded as predicted values, based on input $X = X_i$ for $i = 1, N$, regarded as predictor variable. The dependency between the vectors **X** and **Y** is measured by holding the equality in the following

$$p(x, y) = p(x)p(y) \tag{4}$$

where $p(x)\, and\, p(y)$ are the probability density function of **X** and **Y**, respectively. While $p(x, y)$ is the joint probability density function which characterize the probability distribution of a continuous vectors **X** and **Y**. Equation 4 gives value in range between 0 and 1 for absolute independence and total dependence, respectively. It is useful for detecting the nonlinear dependency between the input and output, in spit the absence of linear dependency between them [5]. Estimating The mutual information for a continuous probability distribution can be performed by the integration of the ratio between the joint probability $p(x, y)$ to the product of $p(x)$ and $p(y)$ [7] as following.

$$R(\boldsymbol{X}, \boldsymbol{Y}) = \int p(x, y) \ln \frac{p(x, y)}{p(x)p(y)} dxdy \tag{5}$$

The integration in Eq. 5 gives value between 0 and $\infty$ and considered as a fundamental role in information theory. To capture the correlation between **X** and **Y** for both linear and nonlinear dependence, one can use the following relation for normalizing the result of Eq. 5 and get the predictability of **Y** from **X**

$$\eta = \sqrt{1 - e^{-2R(\boldsymbol{X}, \boldsymbol{Y})}} \tag{6}$$

Consider the covariance matrix of the random vector $z = (\boldsymbol{X}, \boldsymbol{Y})$ is $\mathcal{C}_z$. For the Gaussian distribution of the joint probability $p(x, y)$, Eq. (5) collapse to some measure of linear dependence as following

$$R(\boldsymbol{X}, \boldsymbol{Y}) = \frac{1}{2} \ln \frac{\det(\mathcal{C}_z)}{\det(\mathcal{C}_x)\det(\mathcal{C}_y)} \tag{7}$$

where $\mathcal{C}_z$ and $\mathcal{C}_y$ are the covariance matrix of **X** and **Y**. This general equation is known as a correlation function that include linear and even nonlinear correlation. Also, it shows that the coefficients of linear correlation define the mutual information between the variables [8]. Nonlinear systems include wide spectra and thus the correlation between variables is not clear and hence the correlation function is not beneficial. The linear predictability of the output by the input can be estimated by the measure of dependency between them as the following

$$\mathcal{L} = \sqrt{1 - \frac{\det(\mathcal{C}_x)\det(\mathcal{C}_y)}{\det(\mathcal{C}_z)}} \tag{8}$$

Obviously, the dependency between the variables can be considered as a powerful tool for estimating the probability of the desired output based on input for any system.

## 5 Conclusion

This chapter discussed the governing equation of satellite orbiting and hazard effects on satellites. It illustrates the different orbits of satellites based on different proposes and showed an overview of the satellite instruments using in remote sensing. It appeared the risk challenges on the satellite's operation and effects of surrounding environments on its behavior like drag effect during its rotation and the threat of satellite loss. In the era of information, the importance of machine learning is shown up to handle all challenges surrounding us. Studying and modelling systems as individual will not reach us to understand well the associations and effects between systems and eventually we can not predict and overcome the challenges. Using the telemetry data as an input in machine learning techniques is crucial for understanding well the behavior of the desired system as well as predicting the anomalies for protecting it during its proposed lifetime. Moreover, it shown at a glance the basic idea of the mutual information theory and how estimating the correlations between variables plays an important role for predicting the probability of occurrence of a specific event. Different tasks of machine learning for working out the facing issues are discussed. This chapter is an introduction of the machine learning applications in remote sensing and telemetry data that can help for presenting an overview of the whole satellite systems and supplement what will not be discussed in the following chapters.

## References


1. S. Liang, X. Li, J. Wang (eds.), *Advanced Remote Sensing: Terrestrial Information Extraction and Applications* (Academic Press, 2012)
2. J. Guo, J.M. Forbes, F. Wei, X. Feng, H. Liu, W. Wan, Z. Yang, C. Liu, B.A. Emery, Y. Deng, Observations of a large-scale gravity wave propagating over an extremely large horizontal distance in the thermosphere. Geophys. Res. Lett. **42**, 6560–6565 (2015)
3. R. Biesbroek, *Active Debris Removal in Space: How to Clean the Earth's Environment from Space Debris* (2015)
4. G. Tóth, I.V. Sokolov, T.I. Gombosi, D.R. Chesney, C.R. Clauer, D.L. DeZeeuw, K.C. Hansen, K.J. Kane, W.B. Manchester, R.C. Oehmke et al., Space weather modeling framework: a new tool for the space science community. J. Geophys. Res. Space Phys. **110**(A12), 226 (2005)
5. N. Gershenfeld, *The Nature of Mathematical Modeling* (Cambridge University Press, Cambridge, 1998)
6. G. Camps-Valls, Machine learning in remote sensing data processing, in *IEEE International Workshop on Machine Learning for Signal Processing, 2009*. MLSP 2009 (IEEE, 2009, September), pp. 1–6
7. A.A. Tsonis, Probing the linearity and nonlinearity in the transitions of the atmospheric circulation. Nonlinear Process. Geophys. **8**, 341–345 (2001)
8. G.A. Darbellay, I. Vajda, Estimation of the information by an adaptive partitioning of the observation space. IEEE Trans. Inf. Theory **45**(4), 1315–1321 (1999)

# Formalization, Prediction and Recognition of Expert Evaluations of Telemetric Data of Artificial Satellites Based on Type-II Fuzzy Sets

**Olga M. Poleshchuk**

**Abstract** Telemetry data from spacecraft or artificial satellites is usually received from numerous sensor outputs connected to various units. These incoming data often contain some symptoms that signal possible fatal system failures. However, the standard methods used cannot always detect these symptoms. The aim of this chapter is to propose formalization, prediction and recognition methods that will help experts to extract important information and not miss the symptoms of possible fatal failures. Investigations of the failures that have already taken place made it possible to draw conclusions about the need to formalize the experience of experts and their knowledge. For the prediction of expert evaluations, regression models based on interval type-II fuzzy sets were developed. The first model is linear and allows predicting expert evaluations of qualitative parameters. The second model is developed for a special class interval type-II fuzzy sets, which can simplify the procedures of expert evaluation. The third model is nonlinear and allows predicting expert evaluations of qualitative parameters. The fourth model with interval type-II fuzzy coefficients is developed for prediction numerical parameters. The methods developed in the chapter open up new possibilities in expert estimation of the parameters of complex objects under conditions of high order uncertainty.

## 1 Introduction

The participation of experts is extremely important when evaluating complex technical objects under conditions of heterogeneous uncertainty. Experts often use verbal scales to assess quantitative and qualitative parameters. The values or levels of these scales are the words of the professional language of experts.

O. M. Poleshchuk (✉)
Space Department, Moscow Bauman State Technical University, Moscow, Russia
e-mail: olga.m.pol@yandex.ru.ru



A. E. Hassanien et al. (eds.), *Machine Learning and Data Mining in Aerospace Technology*, Studies in Computational Intelligence 836,
https://doi.org/10.1007/978-3-030-20212-5_3

The tasks of analyzing the information received from the group of experts and its aggregation are not new, but relevant because with the creation of new, more complex systems in various fields, expert assessment procedures become more complicated and the responsibility of experts increases. This is due to the fact that in conditions of uncertainty of high order and complexity of mathematical modeling, expert estimates are the only data in the evaluation of complex technical systems.

Instruments are used to measure the values of numerical parameters, but experts can also act as measuring instruments, evaluating the parameters in verbal scales. For example, the reaction of a person during an emergency situation is measured in seconds. But experts usually evaluate the human reaction as «very slow», «slow», «normal», «fast» and «very fast». In evaluating the probability of bankruptcy of an enterprise, it is not a numerical value that is important, but an expert's assessment of how high or low this probability is.

In order to associate verbal levels of the scale with the numerical values of the parameter, experts, as a rule, consider the range of parameter values and divide this range into nonintersecting intervals. Each such interval corresponds to a certain level of the verbal scale. The disadvantage of this approach is the lack of smoothness during the transition from one level to another. Features difficulties arise when describing the boundary values. All this together adds uncertainty to the assessment procedure and complicates it.

It is possible to eliminate this disadvantage with the help of fuzzy set theory. In that case we can put fuzzy sets in correspondence with levels of verbal scale, but not intervals [1]. The result of that construction is a linguistic scale. The levels of verbal scale corresponds to physical values of a numerical parameter. The physical values of the numerical parameter are measured by a technical instrument and linguistic values of the numerical parameter are measured by an expert. Each value measured by an instrument belongs to some linguistic value measured by an expert with a degree of expert confidence.

The process of creation a linguistic scale for expert evaluation of a non-numeric parameter is more complicated than creation a linguistic scale for a numerical parameter. This complexity is due to the fact that the qualitative parameter does not have a range of values on the number line. The linguistic scale for a qualitative parameter is a set of verbal values, each of which is associated with a fuzzy set (type-I fuzzy set) [1].

Creation a linguistic scale for expert evaluation of a non-numeric parameter makes it possible to operate correctly with non-comparable values of different qualitative parameters with help of membership functions of their linguistic values.

A fuzzy set $\tilde{A}$ is a pair $\{x, \mu_{\tilde{A}}(x)\}, x \in X$, where $\mu_{\tilde{A}}(x) : X \to [0, 1]$—membership function of $\tilde{A}$, $X$—universal set of $\tilde{A}$ [1].

A set of five $\{X, T(X), U, V, S\}$ was named a linguistic variable, where $T(X) = \{X_l, l = \overline{1, m}\}$—the terms of variable $X$ or names of linguistic values of variable $X$(each of these values is a fuzzy variable with a value from a universal set $U$); $V$—is a syntactical rule that gives names of the values of a linguistic variable $X$;

$S$—is a semantic rule that gives to every fuzzy variable with a name from $T(X)$ a corresponding fuzzy set of $U$ [2].

A linguistic variable with fixed terms was named a semantic scope [2].

The properties of semantic spaces were investigated by a number of authors [3–7]. These studies were aimed at ensuring adequate formalization of the objects under consideration and their heterogeneous characteristics. The research results allowed to formulate requirements for the membership functions $\mu_l(x), l = \overline{1, m}$ of terms $T(X) = \left\{X_l, l = \overline{1, m}\right\}$ of semantic spaces:

1. $\forall X_l, l = \overline{1, m}\ \exists \widehat{U}_l = \{x \in U : \mu_l(x) = 1\} \neq \emptyset$-point or an interval.
2. $\mu_l(x), l = \overline{1, m}$ does not decrease to the left of $\widehat{U}_l$ and does not increase to the right of $\widehat{U}_l$, where $\widehat{U}_l = \{x \in U : \mu_l(x) = 1\}$.
3. $\mu_l(x), l = \overline{1, m}$ have not more two first type points of discontinuity.
4. $\sum\limits_{l=1}^{m} \mu_l(x) = 1\ \forall x \in U$.

It is assumed that each term of a semantic scope has not less one point that belongs to this term with complete expert confidence and each point of $U$ belongs not less to one term of a semantic scope. All the properties 1–4 allow to model the experience of experts and their knowledge. That is why semantic scopes with these properties were often included in intellectual systems of data analysis and decision making [5–8].

The semantic scopes with properties 1–4 were named Full Orthogonal Semantic Scopes (FOSS) [8] and were chosen as expert evaluation models in this chapter.

## 2 Creation of Expert Evaluation Models Based on Type-I Fuzzy Sets

There are some methods for creation FOSS [5, 9, 10] based on different information. It can be information received from an expert or a group of experts. Let us suppose that a verbal scale with levels $X_l, l = \overline{1, m}, m \geq 2$ has been used for evaluation a qualitative parameter $X$ for some $N$ objects. As a result of this we have data of $N$ volume. We formalize received information with help of semantic scope and assign fuzzy numbers $\tilde{X}_l, l = \overline{1, m}$ with membership functions $\mu_l(x), l = \overline{1, m}$ to the levels $X_l, l = \overline{1, m}$. Let we have $n_l, l = \overline{1, m}$ objects with levels $X_l, l = \overline{1, m}$ (accordingly with fuzzy numbers $\tilde{X}_l, l = \overline{1, m}$ and membership functions $\mu_l(x), l = \overline{1, m}$). We denote $\frac{n_l}{N}, l = \overline{1, m}$ by $a_l$, $l = \overline{1, m}$ and $\min(a_l, a_{l+1})$, $l = \overline{1, m-1}$ by $b_l$, $l = \overline{1, m-1}$.

Then according [9, 10]

$$\mu_1(x) \equiv \left(0, a_1 - \frac{b_1}{2}, 0, b_1\right),$$

$$\mu_l(x) \equiv \left(\sum_{i=1}^{l-1} a_i + \frac{b_{l-1}}{2}, \sum_{i=1}^{l} a_i - \frac{b_l}{2}, b_{l-1}, b_l\right), \ l = \overline{2, m-1},$$

$$\mu_m(x) \equiv \left(1 - a_m - \frac{b_{m-1}}{2}, 1 - a_m + \frac{b_{m-1}}{2}, b_{m-1}, 0\right).$$

Fuzzy number with membership function, whose graph is a trapezium is called trapezoidal fuzzy number. Membership function of trapezoidal fuzzy number is defined by four parameters. These parameters are the abscissas of the vertices of the upper base of the trapezium and the length of its wings. Fuzzy number with membership function whose graph is a triangle is called triangular fuzzy number. Membership function of triangular fuzzy number is defined by three parameters. These parameters are the abscissa of the top of the triangle and the length of its wings.

The second method discussed in this chapter is based on expert information regarding points or intervals of the universal set, which with complete expert confidence belong to one or another level of the linguistic scale used.

We construct FOSS on $U = [a, b]$ for a quantitative parameter and on $U = [0, 1]$ for a qualitative parameter. Linguistic scale used has term-set $T(X) = \{X_1, X_2, \ldots, X_m\}$.

An expert supposes intervals $(x_l^1, x_l^2)$ which with complete expert confidence belong to terms $X_l, l = \overline{1, m}$ accordingly.

Membership functions for FOSS are as follows [5]

$$\mu_1(x) \equiv \left(a, x_1^2, 0, \frac{x_2^1 - x_1^2}{2}\right),$$

$$\mu_l(x) \equiv \left(x_l^1, x_l^2, \frac{x_l^1 - x_{l-1}^2}{2}, \frac{x_{l+1}^1 - x_l^2}{2}\right), \; l = \overline{2, m-1},$$

$$\mu_m(x) \equiv \left(x_m^1, b, \frac{x_m^1 - x_{m-1}^2}{2}, 0\right).$$

or

$$\mu_1(x) \equiv \left(0, x_1^2, 0, \frac{x_2^1 - x_1^2}{2}\right),$$

$$\mu_l(x) \equiv \left(x_l^1, x_l^2, \frac{x_l^1 - x_{l-1}^2}{2}, \frac{x_{l+1}^1 - x_l^2}{2}\right), l = \overline{2, m-1},$$

$$\mu_m(x) \equiv \left(x_m^1, 1, \frac{x_m^1 - x_{m-1}^2}{2}, 0\right).$$

## 3 Creation of Generalized Expert Evaluation Models Based on Interval Type-II Fuzzy Sets

Consider $k$ FOSS (expert evaluation models): $X_i = \{\mu_{il}(x), l = \overline{1,m}\}$, $i = \overline{1,k}$, $\mu_{il}(x) \equiv (a_1^{il}, a_2^{il}, a_L^{il}, a_R^{il})$.

In the expert evaluation theory, different indicators are used to calculate the consistency of several expert rankings (the Kendall coefficient [11], the concordance coefficient [4], the rank correlation coefficient in the Kemen-Snell model [11], Spearman's rank correlation coefficient [11]).

In order to analyze expert information, formalized on the basis of linguistic variables, in [5] quantitative indicators of the consistency of expert criteria are determined. For example, the identify index $\kappa$ of the general consistency of $k$ FOSS (expert evaluation models) with membership functions $\{\mu_{il}(x), l = \overline{1,m}\}$, $i = \overline{1,k}$ is determined accordingly as

$$\kappa = \frac{1}{m}\sum_{l=1}^{m} \frac{\int_0^1 \min(\mu_{1l}(x),\ldots,\mu_{kl}(x))dx \int}{\int_0^1 \max(\mu_{1l}(x),\ldots,\mu_{kl}(x))dx},$$
$$0 \le \kappa \le 1.$$

In the expert evaluation theory, the optimality condition of the Pareto group choice is formulated [11]. This condition means that if $R = F(R_1, \ldots, R_k)$—group ranking, which is a function of individual rankings $R_1, \ldots, R_k$, then $\cap_{n=1}^{k} R_n \subseteq R \subseteq \cup_{n=1}^{k} R_n$.

Let us determine a generalized expert model $X$ (with membership functions $\{f_l(x), l = \overline{1,m}\}$, $f_l(x) \equiv (a_1^l, a_2^l, a_L^l, a_R^l)$) based on expert evaluation models $X_i = \{\mu_{il}(x), l = \overline{1,m}\}$, $i = \overline{1,k}$, $\mu_{il}(x) \equiv (a_1^{il}, a_2^{il}, a_L^{il}, a_R^{il})$ with weight coefficients $\omega_i$, $i = \overline{1,k}$ and formulate the condition of the Pareto group choice:

$$\bigcap_{i=1}^{k} X_i \subseteq X \subseteq \bigcup_{i=1}^{k} X_i$$

or

$$\min_x(\mu_{1l}(x), \ldots, \mu_{kl}(x)) \le f_l(x) \le$$
$$\le \max_x(\mu_{1l}(x), \ldots, \mu_{kl}(x))$$
$$\forall l = \overline{1,m}, x \in [0, 1].$$

Unknown parameters $a_1^l, a_2^l, a_L^l, a_R^l$, $l = \overline{1,m}$ are determined from the condition:

$$F = \sum_{j=1}^{m}\sum_{i=1}^{k} \omega_i \Big[ (a_1^{il} - a_1^l)^2 + (a_2^{il} - a_2^l)^2 +$$
$$+ (a_L^{il} - a_L^l)^2 + (a_R^{il} - a_R^l)^2 \Big] \to \min,$$

Unknown parameters $a_1^l, a_2^l, a_L^l, a_R^l,\ l = \overline{1, m}$ of generalized expert model are as follows:

$$a_1^l = \sum_{i=1}^{k} \omega_i a_1^{il}, l = \overline{1, m},$$
$$a_2^l = \sum_{i=1}^{k} \omega_i a_2^{il}, l = \overline{1, m},$$
$$a_L^l = \sum_{i=1}^{k} \omega_i a_L^{il}, l = \overline{1, m},$$
$$a_R^l = \sum_{i=1}^{k} \omega_i a_R^{il}, l = \overline{1, m}.$$

Constructed generalized expert model is satisfied the condition of the Pareto group choice [5].

However, it should be noted that the generalized model obtained is a kind of average opinion of different experts. This should be attributed to the minuses rather than the pluses, because we would like to get not only an average opinion, but the spread of expert criterions and the fuzziness of the degree of expert confidence in evaluating a particular parameter.

Type-II fuzzy sets can help in that [12].

A type-II fuzzy set is $\{x, \mu_{\tilde{A}}(x)\}, x \in X$, where the value of membership function $\mu_{\tilde{A}}(x)$ is a type-I fuzzy set [12].

An interval type-II fuzzy set is defined by low membership function and upper membership function [11], which are denoted by $\underline{\mu_{\tilde{A}}}$ and $\overline{\mu_{\tilde{A}}}$ respectively (Fig. 1), $\underline{\mu_{\tilde{A}}} = \left(a_1^L, a_2^L,\ a_l^L, a_r^L\right), \overline{\mu_{\tilde{A}}} = \left(a_1^U, a_2^U, a_l^U, a_r^U\right)$.

Interval type-II fuzzy sets allow to save individual expert information about a word and use this information to get a generalized expert model.

Let us consider parameters $a_1^{il}, a_2^{il}, i = \overline{1, k},\ l = \overline{1, m}$ of $k$ expert evaluation models:

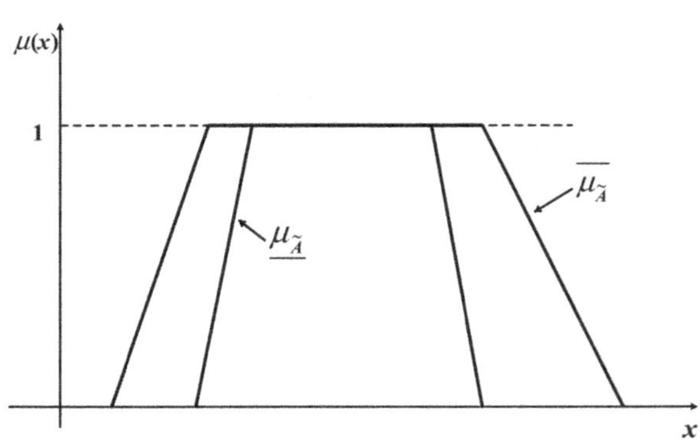


**Fig. 1** Interval type-II fuzzy set $\tilde{A}$ with $\underline{\mu_{\tilde{A}}}$ and $\overline{\mu_{\tilde{A}}}$

$$a_1^l = \sum_{i=1}^{k} \omega_i a_1^{il}, l = \overline{1, m},$$

$$a_2^l = \sum_{i=1}^{k} \omega_i a_2^{il}, l = \overline{1, m}.$$

Let us calculate $s_{1l}^2 = \frac{1}{k-1}\sum_{i=1}^{k}\left(a_1^{il} - a_1^l\right)^2$ и $s_{2l}^2 = \frac{1}{k-1}\sum_{i=1}^{k}\left(a_2^{il} - a_2^l\right)^2, l = \overline{1, m}$ and construct confidence intervals for the parameters $\hat{a}_1^l, \hat{a}_2^l, l = \overline{1, m}$ of the generalized expert model, using the Student distribution

$$\sum_{i=1}^{k} \omega_i a_1^{il} - \frac{s_{1l}\Delta_{k-1,\alpha}}{\sqrt{k}} \le \hat{a}_1^l \le \sum_{i=1}^{k} \omega_i a_1^{il} + \frac{s_{1l}\Delta_{k-1,\alpha}}{\sqrt{k}},$$
$$l = \overline{1, m}$$

$$\sum_{i=1}^{k} \omega_i a_2^{il} - \frac{s_{2l}\Delta_{k-1,\alpha}}{\sqrt{k}} \le \hat{a}_2^l \le \sum_{i=1}^{k} \omega_i a_2^{il} + \frac{s_{2l}\Delta_{k-1,\alpha}}{\sqrt{k}}, l = \overline{1, m},$$

where $\Delta_{k-1,\alpha}$ is found from the table for probabilities $P\left(|t_{k-1}| > \Delta_{k-1,\alpha}\right) = \alpha$ of Student's distribution $t_{k-1}t_{k-1}$.

Proceeding from this, we present a generalized expert model in the form of a linguistic variable, whose values are interval type-II fuzzy sets, the upper $\overline{f_l(x)}$ and lower $\underline{f_l(x)}$ membership functions of which are respectively specified by parameters:

$$\overline{f_l(x)} = \left(\sum_{i=1}^{k} \omega_i a_1^{il} - \frac{s_{1l}\Delta_{k-1,\alpha}}{\sqrt{k}}, \sum_{i=1}^{k} \omega_i a_2^{il} + \frac{s_{2l}\Delta_{k-1,\alpha}}{\sqrt{k}},\right.$$
$$\left.\sum_{i=1}^{k} \omega_i a_L^{il}, \sum_{i=1}^{k} \omega_i a_R^{il}\right), l = \overline{1, m},$$

$$\underline{f_l(x)} = \left(\sum_{i=1}^{k} \omega_i a_1^{il} + \frac{s_{1l}\Delta_{k-1,\alpha}}{\sqrt{k}}, \sum_{i=1}^{k} \omega_i a_2^{il} - \frac{s_{2l}\Delta_{k-1,\alpha}}{\sqrt{k}},\right.$$
$$\left.\sum_{i=1}^{k} \omega_i a_L^{il}, \sum_{i=1}^{k} \omega_i a_R^{il}\right), l = \overline{1, m}.$$

If $\sum_{i=1}^{k} \omega_i a_1^{il} + \frac{s_{1l}\Delta_{k-1,\alpha}}{\sqrt{k}} > \sum_{i=1}^{k} \omega_i a_2^{il}$, then

$$\overline{f_l(x)} = \left(\sum_{i=1}^{k} \omega_i a_1^{il} - \frac{s_{1l}\Delta_{k-1,\alpha}}{\sqrt{k}}, \sum_{i=1}^{k} \omega_i a_2^{il} + \frac{s_{2l}\Delta_{k-1,\alpha}}{\sqrt{k}},\right.$$
$$\left.\sum_{i=1}^{k} \omega_i a_L^{il}, \sum_{i=1}^{k} \omega_i a_R^{il}\right), l = \overline{1, m},$$

$$\underline{f_l(x)} = \left(\sum_{i=1}^{k} \omega_i a_1^{il}, \sum_{i=1}^{k} \omega_i a_2^{il} - \frac{s_{2l}\Delta_{k-1,\alpha}}{\sqrt{k}},\right.$$
$$\left.\sum_{i=1}^{k} \omega_i a_L^{il}, \sum_{i=1}^{k} \omega_i a_R^{il}\right), l = \overline{1, m}.$$

If $\sum_{i=1}^{k} \omega_i a_2^{il} - \frac{s_{2l}\Delta_{k-1,\alpha}}{\sqrt{k}} < \sum_{i=1}^{k} \omega_i a_1^{il}$, then

$$\overline{f_l(x)} = \left(\sum_{i=1}^{k} \omega_i a_1^{il} - \frac{s_{1l}\Delta_{k-1,\alpha}}{\sqrt{k}}, \sum_{i=1}^{k} \omega_i a_2^{il} + \frac{s_{2l}\Delta_{k-1,\alpha}}{\sqrt{k}},\right.$$
$$\left.\sum_{i=1}^{k} \omega_i a_L^{il}, \sum_{i=1}^{k} \omega_i a_R^{il}\right), l = \overline{1, m},$$
$$\underline{f_l(x)} = \left(\sum_{i=1}^{k} \omega_i a_1^{il} + \frac{s_{1l}\Delta_{k-1,\alpha}}{\sqrt{k}}, \sum_{i=1}^{k} \omega_i a_2^{il},\right.$$
$$\left.\sum_{i=1}^{k} \omega_i a_L^{il}, \sum_{i=1}^{k} \omega_i a_R^{il}\right), l = \overline{1, m}.$$

If $\sum_{i=1}^{k} \omega_i a_1^{il} + \frac{s_{1l}\Delta_{k-1,\alpha}}{\sqrt{k}} > \sum_{i=1}^{k} \omega_i a_2^{il}$ and $\sum_{i=1}^{k} \omega_i a_2^{il} - \frac{s_{2l}\Delta_{k-1,\alpha}}{\sqrt{k}} < \sum_{i=1}^{k} \omega_i a_1^{il}$, then

$$\overline{f_l(x)} = \left(\sum_{i=1}^{k} \omega_i a_1^{il} - \frac{s_{1l}\Delta_{k-1,\alpha}}{\sqrt{k}}, \sum_{i=1}^{k} \omega_i a_2^{il} + \frac{s_{2l}\Delta_{k-1,\alpha}}{\sqrt{k}},\right.$$
$$\left.\sum_{i=1}^{k} \omega_i a_L^{il}, \sum_{i=1}^{k} \omega_i a_R^{il}\right), l = \overline{1, m},$$
$$\underline{f_l(x)} = \left(\sum_{i=1}^{k} \omega_i a_1^{il}, \sum_{i=1}^{k} \omega_i a_2^{il},\right.$$
$$\left.\sum_{i=1}^{k} \omega_i a_L^{il}, \sum_{i=1}^{k} \omega_i a_R^{il}\right), l = \overline{1, m}.$$

For the upper membership function of the first term, the first parameter is assumed to be zero, for the upper membership function of the last term, the second parameter is set equal to one.

The generalized expert model in the form of a linguistic variable, whose values are interval type-II fuzzy sets, takes into account the scatter of expert opinions and allows to obtain an interval estimation of the degree of confidence of a group of experts in a particular solution.

## 4 Weighted Intervals for Interval Type-2 Fuzzy Sets

Let us consider a interval type-II fuzzy set $\tilde{A}$ defined by low membership function and upper membership function, which are denoted by $\underline{\mu_{\tilde{A}}}$ and $\overline{\mu_{\tilde{A}}}$ respectively, $\underline{\mu_{\tilde{A}}} = \left(a_1^L, a_2^L,\ a_l^L, a_r^L\right), \overline{\mu_{\tilde{A}}} = \left(a_1^U, a_2^U, a_l^U, a_r^U\right)$.

The definition of weighted point $B$ for a triangular number $\tilde{B} = (b, b_l, b_r)$ was given in [13]:

$$B = \frac{\int_0^1 \left(\frac{B_\alpha^1 + B_\alpha^2}{2}\right) 2\alpha d\alpha}{\int_0^1 2\alpha d\alpha} = \int_0^1 (B_\alpha^1 + B_\alpha^2)\alpha d\alpha =$$

$$= \int_0^1 (2b + (1-\alpha)b_r - (1-\alpha)b_l)\alpha d\alpha = b + \frac{1}{6}(b_r - b_l).$$

When finding the weighted point the definition of $\alpha$-cut for a fuzzy number $\tilde{B} = (b, b_l, b_r)$ has been used.
$\alpha$-cut of fuzzy number $\tilde{B} = (b, b_l, b_r)$ is interval $\left[B_\alpha^1, B_\alpha^2\right]$, where $B_\alpha^1 = b - (1-\alpha)b_l$, $B_\alpha^2 = b + (1-\alpha)b_r$.

According to this definition two triangular numbers with different second and third parameters have the same weighted points. Let. $\tilde{A} = (2, 3, 3)$, $\tilde{B} = (2, 4, 4)$. We calculate the weighted points $A, B$ for numbers $\tilde{A}, \tilde{B}$ accordingly as follows:

$$A = \int_0^1 (4 - 3(1-\alpha) + 3(1-\alpha))\alpha d\alpha = 2,$$

$$B = \int_0^1 (4 - 4(1-\alpha) + 4(1-\alpha))\alpha d\alpha = 2.$$

Sometimes this is not a problem to solve some practical tasks. But for other tasks this is may a problem, for example, when it is necessity to accumulate more information about input fuzzy numbers and to save their properties in an aggregative indicator. It is especially important in decision-making.

To eliminate this lack of a weighted point we propose the definition of a weighted interval.

For a start we will define the weighted set for the trapezoidal fuzzy number $\tilde{A} \equiv \left(a_1, a_2, a_l, a_r\right)$ as the set of weighted points of all triangular numbers $\tilde{B} \equiv (b, b_l, b_r)$ that belong to the number $\tilde{A}$ [14, 15].

**Proposition 1** [5] *The weighted set for the trapezoidal fuzzy number* $\tilde{A} = \left(a_1, a_2, a_l, a_r\right)$ *is an interval* $[A_1, A_2]$, *such as*

$$A_1 = a_1 - \frac{1}{6}a_l, A_2 = a_2 + \frac{1}{6}a_r.$$

We shall call the interval $[A_1, A_2]$ the weighted interval for trapezoidal fuzzy number $\tilde{A} = (a_1, a_{2,} a_l, a_r)$.

Let us consider two triangular fuzzy numbers: $\tilde{A} \equiv (2, 3, 3)$, $\tilde{B} \equiv (2, 4, 4)$ again and define the weighted intervals $[A_1, A_2]$, $[B_1, B_2]$ for numbers $\tilde{A}$, $\tilde{B}$.

$$A_1 = \int_0^1 (4 - 3(1 - \alpha))\alpha d\alpha = 2 - 3 \times \frac{1}{6} = 1\frac{1}{2},$$

$$A_2 = \int_0^1 (4 + 3(1 - \alpha))\alpha d\alpha = 2 + 3 \times \frac{1}{6} = 2\frac{1}{2},$$

$$B_1 = \int_0^1 (4 - 4(1 - \alpha))\alpha d\alpha = 2 - 4 \times \frac{1}{6} = 1\frac{1}{3},$$

$$B_2 = \int_0^1 (4 + 4(1 - \alpha))\alpha d\alpha = 2 + 4 \times \frac{1}{6} = 2\frac{2}{3},$$

$$[A_1, A_2] = \left[1\frac{1}{2}, 2\frac{1}{2}\right], [B_1, B_2] = \left[1\frac{1}{3}, 2\frac{2}{3}\right].$$

We can see that fuzzy numbers $\tilde{A} \equiv (2, 3, 3)$, $\tilde{B} \equiv (2, 4, 4)$ have the same weighted points but the different weighted intervals.

**Proposition 2** [5] *If* $[A_1, A_2]$, $[B_1, B_2]$ *are the weighted intervals for fuzzy numbers* $\tilde{A}$, $\tilde{B}$ *then* $[A_1 + B_1, A_2 + B_2]$ *is the weighted interval for fuzzy number* $\tilde{A} + \tilde{B}$.

**Proposition 3** [5] *The weighted interval for the number* $\tilde{D} = \tilde{A} \times \tilde{B}$ *is defined by linear combinations of parameters* $\tilde{A}$, $\tilde{B}$.

Let us consider $\tilde{A} = (a_1, a_{2,} a_l, a_r) \geq 0$ and a triangular number $\tilde{a} \equiv (b, b_l, b_r)$.

**Proposition 4** [5] *The weighed interval* $\left[\theta^1_{\tilde{a}\tilde{A}}, \theta^2_{\tilde{a}\tilde{A}}\right]$ *for the number* $\tilde{a} \times \tilde{A}$ *is defined as follows*

$$\theta^1_{\tilde{a}\tilde{A}} = b\left(a_q + (-1)^q \frac{1}{6}a_{M_q}\right) - b_l\left(\frac{1}{6}a_q + (-1)^q \frac{1}{12}a_{M_q}\right).$$

$$\theta^2_{\tilde{a}\tilde{A}} = b\left(a_p + (-1)^p \frac{1}{6}a_{M_p}\right) + b_r\left(\frac{1}{6}a_p + (-1)^q \frac{1}{12}a_{M_p}\right).$$

$$q = \begin{cases} 1, b - b_l \geq 0 \\ 2, b + b_r < 0 \end{cases}, M_q = \begin{cases} l, q = 1 \\ r, q = 2 \end{cases},$$

$$p = \begin{cases} 2, \; b - b_l \geq 0 \\ 1, \; b + b_r < 0 \end{cases}, \; M_p = \begin{cases} l, \; p = 1 \\ r, \; p = 2 \end{cases}.$$

**Proposition 5** [14] *The weighed interval* $\left[\theta^1_{\tilde{a}\tilde{A}^2}, \theta^2_{\tilde{a}\tilde{A}^2}\right]$ *for the number* $\tilde{a} \times \tilde{A}^2$ *is defined as follows*

$$\theta^1_{\tilde{a}\tilde{A}^2} = b\left[a_q^2 + \frac{(-1)^q}{3} a_q a_{M_q} + \frac{1}{12} a_{M_q}^2\right] - \\ - b_L\left[\frac{1}{6} a_q^2 + \frac{(-1)^q}{6} a_q a_{M_q} + \frac{1}{6} a_{M_q}^2\right];$$

$$\theta_{\tilde{a}\tilde{A}^2} = b\left[a_r^2 + \frac{(-1)^r}{3} a_r a_{M_r} + \frac{1}{12} a_{M_r}^2\right] + \\ + b_R\left[\frac{1}{6} a_r^2 + \frac{(-1)^r}{6} a_r a_{M_r} + \frac{1}{20} a_{M_r}^2\right].$$

Let determine aggregation intervals $\left[A_1^L, A_2^L\right], \left[A_1^U, A_2^U\right]$ for low membership function $\underline{\mu_{\tilde{A}}} = \left(a_1^L, a_2^L, a_l^L, a_r^L\right)$ and upper membership function $\overline{\mu_{\tilde{A}}} = \left(a_1^U, a_2^U, a_l^U, a_r^U\right)$ of interval type-II fuzzy set $\tilde{A}$:

$$A_1^L = a_1^L - \frac{1}{6} a_l^L, \; A_2^L = a_2^L + \frac{1}{6} a_r^L,$$

$$A_1^U = a_1^U - \frac{1}{6} a_l^U, \; A_2^U = a_2^U + \frac{1}{6} a_r^U.$$

Let $f^2\left(\tilde{A}, \tilde{B}\right) = \left(A_1^L - B_1^L\right)^2 + \left(A_2^L - B_2^L\right)^2 + \left(A_1^U - B_1^U\right)^2 + \left(A_2^U - B_2^U\right)^2$, where $\left[A_1^L, A_2^L\right], \left[A_1^U, A_2^U\right], \left[B_1^L, B_2^L\right], \left[B_1^U, B_2^U\right]$ are weighed intervals for fuzzy sets $\tilde{A}, \tilde{B}$.

## 5 Prediction of Expert Evaluations Based on Linear Regression with Initial Interval Type-II Data

Consider the main approaches to the construction of fuzzy regressions.

(a) The first regression model has been developed by Tanaka [16]. Based on this model other models have been developed [17–25]. The basis of these models is the theory of possibilities instead of probability theory or the both theories. The coefficients of these models are triangular numbers.

(b) In [26] another approach is proposed for the construction of a regression model. This approach is based on fuzzy c-means clustering algorithm. An ordinary regression is constructed for each fuzzy cluster. When all the regressions constructed the most appropriate regression is determined and this regression is used for a new input.

(c) In [27–29] a different approach has been developed for constructing fuzzy regression models based on fuzzy functions and fuzzy c-means clustering algorithm.

All described approaches consider only type-I fuzzy sets, which significantly limits the scope of the developed fuzzy regression models. Experts often use professional language words that can be formalized based on type-II fuzzy sets. However, in the described approaches this possibility is not provided. It is more difficult to operate with type-II fuzzy sets than with type-I fuzzy sets. Perhaps this complexity explains the long absence of regression models based on type-II fuzzy sets. We will consider interval type-II fuzzy sets. To simplify the task of operating with interval type-II fuzzy sets will allow a new concept of weighted interval. This concept underlies the method for constructing fuzzy regression models. The main idea of the method is to determine weighted intervals for low membership function and upper membership function of interval type-II fuzzy set.

Let $\tilde{Y}_i i = \overline{1,n}$ are output interval type-II fuzzy sets, defined by low membership functions $\mu_{\tilde{Y}_i} = (y_1^{iL}, y_2^{iL}, y_l^{iL}, y_r^{iL})$, $i = \overline{1,n}$ and upper membership functions $\overline{\mu_{\tilde{Y}_i}} = (y_1^{iU}, y_2^{iU}, y_l^{iU}, y_r^{iU})$, $y_1^{iU} - y_l^{iU} \geq 0\ i = \overline{1,n}$.

Let $\tilde{X}_j^i$, $j = \overline{1,m}$, $i = \overline{1,n}$ input interval type-II fuzzy sets, defined by low membership functions $\mu_{\tilde{X}_j^i} = \left(x_1^{jiL}, x_2^{jiL}, x_l^{jiL}, x_r^{jiL}\right)$ and upper membership functions $\overline{\mu_{\tilde{X}_j^i}} = \left(x_1^{jiU}, x_2^{jiU}, x_l^{jiU}, x_r^{jiU}\right)$, $x_1^{jiU} - x_l^{jiU} \geq 0$, $j = \overline{1,m}$, $i = \overline{1,n}$.

We will construct a fuzzy regression model as follows:

$$\tilde{Y} = \tilde{a}_0 + \tilde{a}_1\tilde{X}_1 + \cdots + \tilde{a}_m\tilde{X}_m,$$

where $\tilde{a}_j \equiv \left(b^j, b_l^j, b_r^j\right)$, $j = \overline{0,m}$ are triangular fuzzy numbers.

Construction of regression model is carried out using weighted intervals for low membership function and upper membership function of interval type-II fuzzy set.

We determine the weighed intervals $\left[\theta_{\hat{Y}_i}^{1L}, \theta_{\hat{Y}_i}^{2L}\right]$, $\left[\theta_{\hat{Y}_i}^{1U}, \theta_{\hat{Y}_i}^{2U}\right]$ for $\hat{Y}_i = \tilde{a}_0 + \tilde{a}_1\tilde{X}_1^i + \cdots + \tilde{a}_m\tilde{X}_m^i$ using Propositions 1–4:

$$\theta_{\hat{Y}_i}^{1L} = b^0 - \frac{1}{6}b_l^0 + \sum_{j=1}^{m} \theta_{\tilde{a}_j\tilde{X}_j^i}^{1L}\left(b^j, b_l^j, b_r^j\right),$$

$$\theta_{\hat{Y}_i}^{2L} = b^0 - \frac{1}{6}b_l^0 + \sum_{j=1}^{m} \theta_{\tilde{a}_j\tilde{X}_j^i}^{2L}\left(b^j, b_l^j, b_r^j\right),$$

$$\theta_{\hat{Y}_i}^{1U} = b^0 - \frac{1}{6}b_l^0 + \sum_{j=1}^{m} \theta_{\tilde{a}_j\tilde{X}_j^i}^{1U}\left(b^j, b_l^j, b_r^j\right),$$

$$\theta_{\hat{Y}_i}^{2U} = b^0 - \frac{1}{6}b_l^0 + \sum_{j=1}^{m} \theta_{\tilde{a}_j\tilde{X}_j^i}^{2U}\left(b^j, b_l^j, b_r^j\right).$$

$$\theta^{1L}_{\tilde{a}_j\tilde{X}^i_j}\left(b^j, b^j_l, b^j_r\right) = b^j\left(x^{jiL}_q + (-1)^q\frac{1}{6}x^{jiL}_{M_q}\right) - b^j_l\left(\frac{1}{6}x^{jiL}_q + (-1)^q\frac{1}{12}x^{jiL}_{M_q}\right).$$

$$\theta^{2L}_{\tilde{a}_j\tilde{X}^i_j}\left(b^j, b^j_l, b^j_r\right) = b^j\left(x^{jiL}_p + (-1)^p\frac{1}{6}x^{jiL}_{M_p}\right) + b^j_r\left(\frac{1}{6}x^{jiL}_p + (-1)^q\frac{1}{12}x^{jiL}_{M_p}\right).$$

$$\theta^{1U}_{\tilde{a}_j\tilde{X}^i_j}\left(b^j, b^j_l, b^j_r\right) = b^j\left(x^{jiU}_q + (-1)^q\frac{1}{6}x^{jiU}_{M_q}\right) - b^j_l\left(\frac{1}{6}x^{jiU}_q + (-1)^q\frac{1}{12}x^{jiU}_{M_q}\right).$$

$$\theta^{2U}_{\tilde{a}_j\tilde{X}^i_j}\left(b^j, b^j_l, b^j_r\right) = b^j\left(x^{jiU}_p + (-1)^p\frac{1}{6}x^{jiU}_{M_p}\right) + b^j_r\left(\frac{1}{6}x^{jiU}_p + (-1)^q\frac{1}{12}x^{jiU}_{M_p}\right),$$

$$q = \begin{cases} 1, b - b_l \geq 0 \\ 2, b + b_r < 0 \end{cases}, M_q = \begin{cases} l, q = 1 \\ r, q = 2 \end{cases}, p = \begin{cases} 2, b - b_l \geq 0 \\ 1, b + b_r < 0 \end{cases}, M_p = \begin{cases} l, p = 1 \\ r, p = 2 \end{cases}.$$

It is complicated enough to determine low membership functions and upper membership functions of model output data because it is not always trapezoidal fuzzy numbers. That is why we will use the definition of $\alpha$-cuts for a fuzzy number.

If $\tilde{a} \equiv (b, b_l, b_r) < 0$, $\tilde{A} = (a_1, a_2, a_l, a_r) \geq 0$ then $\alpha$-cut $\left[C^1_\alpha, C^2_\alpha\right]$ of $\tilde{a}\tilde{A}$ looks like:

$$C^1_\alpha = ba_2 + (1-\alpha)ba_r - (1-\alpha)b_l a_2 - (1-\alpha)^2 b_l a_r,$$

$$C^2_\alpha = ba_1 + (1-\alpha)ba_l + (1-\alpha)b_r a_1 - (1-\alpha)^2 b_r a_l.$$

If $\tilde{a} \equiv (b, b_l, b_r) \geq 0$, $\tilde{A} = (a_1, a_2, a_l, a_r) \geq 0$ then $\alpha$-cut $\left[C^1_\alpha, C^2_\alpha\right]$ of $\tilde{a}\tilde{A}$ looks like:

$$C^1_\alpha = ba_1 - (1-\alpha)ba_l - (1-\alpha)b_r a_1 + (1-\alpha)^2 b_r a_l,$$

$$C^2_\alpha = ba_2 + (1-\alpha)ba_r + (1-\alpha)b_l a_2 + (1-\alpha)^2 b_l a_r.$$

According to the definition of weighed interval we get weighed intervals $\left[\theta^{1L}_{\tilde{Y}_i}, \theta^{2L}_{\tilde{Y}_i}\right], \left[\theta^{1U}_{\tilde{Y}_i}, \theta^{2U}_{\tilde{Y}_i}\right]$ for initial output data $\tilde{Y}_i i = \overline{1, n}$:

$$\theta^{1L}_{\tilde{Y}_i} = y^{iL}_1 - \frac{1}{6}y^{iL}_l, \theta^{2L}_{\tilde{Y}_i} = y^{iL}_2 + \frac{1}{6}y^{iL}_r,$$

$$\theta^{1U}_{\tilde{Y}_i} = y^{iU}_1 - \frac{1}{6}y^{iU}_l, \theta^{2U}_{\tilde{Y}_i} = y^{iU}_2 + \frac{1}{6}y^{iU}_r.$$

We determine a functional

$$F\left(b^j, b^j_l, b^j_r\right) = \sum_{i=1}^{n} f^2\left(\hat{Y}_i, \tilde{Y}_i\right) = \sum_{i=1}^{n}\left[\left(\theta^{1L}_{\hat{Y}_i} - \theta^{1L}_{\tilde{Y}_i}\right)^2 + \left(\theta^{2L}_{\hat{Y}_i} - \theta^{2L}_{\tilde{Y}_i}\right)^2\right] + \sum_{i=1}^{n}\left[\left(\theta^{1U}_{\hat{Y}_i} - \theta^{1U}_{\tilde{Y}_i}\right)^2 + \left(\theta^{2U}_{\hat{Y}_i} - \theta^{2U}_{\tilde{Y}_i}\right)^2\right],$$

then

$$F\left(b^j, b_l^j, b_r^j\right) = \sum_{i=1}^{n}\left[b^0 - \frac{1}{6}b_l^0 - y_1^{iL} + \frac{1}{6}y_l^{iL} + \sum_{j=1}^{m}\theta_{\tilde{a}_j\tilde{X}_j^i}^{1L}\left(b^j, b_l^j, b_r^j\right)\right]^2 +$$
$$+\sum_{i=1}^{n}\left[b^0 + \frac{1}{6}b_r^0 - y_2^{iL} - \frac{1}{6}y_r^{iL} + \sum_{j=1}^{m}\theta_{\tilde{a}_j\tilde{X}_j^i}^{2L}\left(b^j, b_l^j, b_r^j\right)\right]^2 +$$
$$+\sum_{i=1}^{n}\left[b^0 - \frac{1}{6}b_l^0 - y_1^{iU} + \frac{1}{6}y_l^{iU} + \sum_{j=1}^{m}\theta_{\tilde{a}_j\tilde{X}_j^i}^{1U}\left(b^j, b_l^j, b_r^j\right)\right]^2 +$$
$$+\sum_{i=1}^{n}\left[b^0 + \frac{1}{6}b_r^0 - y_2^{iU} - \frac{1}{6}y_r^{iU} + \sum_{j=1}^{m}\theta_{\tilde{a}_j\tilde{X}_j^i}^{2U}\left(b^j, b_l^j, b_r^j\right)\right]^2.$$

It is easy to see that $F\left(b^j, b_l^j, b_r^j\right)$ is piecewise differentiable function in the field $b_l^j \geq 0,\ b_r^j \geq 0,\ j = \overline{0, m}$ because $\theta_{\tilde{a}_j\tilde{X}_j^i}^{1L}\left(b^j, b_l^j, b_r^j\right)$, $\theta_{\tilde{a}_j\tilde{X}_j^i}^{2L}\left(b^j, b_l^j, b_r^j\right)$, $\theta_{\tilde{a}_j\tilde{X}_j^i}^{1U}\left(b^j, b_l^j, b_r^j\right)$, $\theta_{\tilde{a}_j\tilde{X}_j^i}^{2U}\left(b^j, b_l^j, b_r^j\right)$ are piecewise linear functions.

We will find unknown parameters from the condition:

$$F\left(b^j, b_l^j, b_r^j\right) = \sum_{i=1}^{n} f^2\left(\hat{Y}_i, \tilde{Y}_i\right) \to \min,$$
$$b_l^j \geq 0, b_r^j \geq 0,\ j = \overline{0, m}$$

by known methods [30].

The quality indicators of the regression model play a significant role. By analogy with the classical regression model, we define the standard deviation of the output variable ($S_{\tilde{y}}$), the correlation coefficient ($HR^2$) and the standard error of estimates of the output variable ($HS$) [31, 32]:

$$S_{\tilde{y}} = \sqrt{\frac{1}{n-1}\sum_{i=1}^{n} f^2\left(\tilde{Y}_i, \bar{\tilde{Y}}\right)},\ \bar{\tilde{Y}} = \frac{\sum_{i=1}^{n}\tilde{Y}_i}{n},$$
$$HR^2 = \frac{\sum_{i=1}^{n} f^2\left(\hat{Y}_i, \bar{\tilde{Y}}\right)}{\sum_{i=1}^{n} f^2\left(\tilde{Y}_i, \bar{\tilde{Y}}\right)},\ HS = \sqrt{\frac{1}{n-m-1}\sum_{i=1}^{n} f^2\left(\hat{Y}_i, \tilde{Y}_i\right)}.$$

Suppose that for evaluation output parameter $Y$ experts use a linguistic scale with levels $Y_k,\ k = \overline{1, p}$, that are formalized with the help of interval type-II fuzzy sets $\tilde{\tilde{Y}}_k,\ k = \overline{1, p}$ defined by their low membership functions $\underline{\mu_{\tilde{\tilde{Y}}_k}} = \left(y_1^{kL}, y_2^{kL}, y_l^{kL}, y_r^{kL}\right), k = \overline{1, p}$ and upper membership functions $\overline{\mu_{\tilde{\tilde{Y}}_k}} =$

$\left(y_1^{kU}, y_2^{kU}, y_l^{kU}, y_r^{kU}\right), k = \overline{1, p}$. We have got model output value $\hat{Y}_i$ of regression model in the form of interval type-II fuzzy set. But it is very important to identify this fuzzy set with one of the levels $Y_k$, $k = \overline{1, p}$ of the linguistic scale used by experts [33].

Let $\left[C_1^{iL}, C_2^{iL}\right]$, $\left[C_1^{iL}, C_2^{iL}\right]$, $i = \overline{1, n}$ are weighted intervals of $\hat{Y}_i$, $\left[D_1^{iL}, D_2^{iL}\right]$, $\left[D_1^{iL}, D_2^{iL}\right]$, $k = \overline{1, p}$ are weighted intervals of $\tilde{\tilde{Y}}_k$, $k = \overline{1, p}$.

Then

$$f^2\left(\hat{Y}_i, \tilde{\tilde{Y}}_k\right) = \left(C_1^{iL} - D_1^{kL}\right)^2 + \left(C_2^{iL} - D_2^{kL}\right)^2 + \\ + \left(C_1^{iU} - D_1^{iU}\right)^2 + \left(C_2^{iU} - D_2^{iU}\right)^2, i = \overline{1, n}, k = \overline{1, p}.$$

The value of regression model $\hat{Y}_i$ is identified to level $Y_s$ of the linguistic scale used, if

$$f^2\left(\hat{Y}_i, \tilde{\tilde{Y}}_s\right) = \min_k f^2\left(\hat{Y}_i, \tilde{\tilde{Y}}_k\right), k = \overline{1, p}.$$

## 6 Prediction of Expert Evaluations Based on Linear Regression with Initial Special Case of Interval Type-2 Fuzzy Sets

In this section we consider fuzzy sets, one of which $\tilde{A}$, as a typical representative, is shown in Fig. 2.

This fuzzy set is defined by low membership function $\underline{\mu_{\tilde{A}}} = \left(a^L, a_l^L, a_r^L\right)$ and upper membership function $\overline{\mu_{\tilde{A}}} = \left(a^U, a_l^U, a_r^U\right)$. Membership function of triangular fuzzy number is defined by three parameters. These parameters are the abscissa of the top of the triangle and the length of its wings.

Let us consider nonnegative $\tilde{A} \equiv (a, a_l, a_r)$ and $\tilde{a} \equiv \left(b, b_l, b_r\right)$.

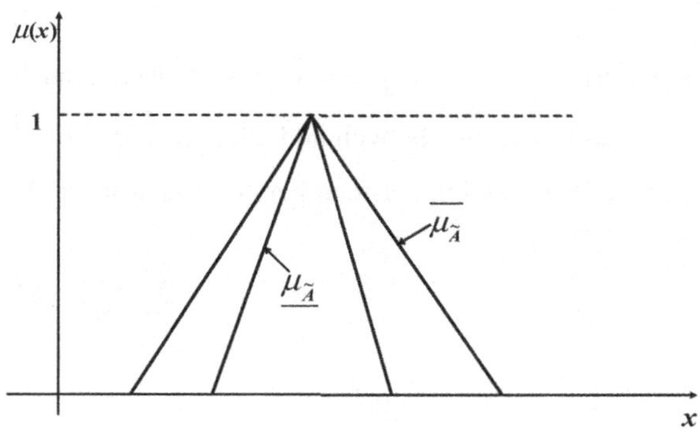


**Fig. 2** Interval type-II fuzzy set $\tilde{A}$ with LMF $\underline{\mu_{\tilde{A}}}$ and UMF $\overline{\mu_{\tilde{A}}}$

Boundaries of the weighed interval $\left[\theta^1_{\tilde{a}\tilde{A}},\ \theta^2_{\tilde{a}\tilde{A}}\right]$ of product of fuzzy numbers $\tilde{a}$ and $\tilde{A}$ look like [34]

$$\theta^1_{\tilde{a}\tilde{A}} = b\left(a + (-1)^q \frac{1}{6} a_{M_q}\right) - b_l\left(\frac{1}{6}a + (-1)^q \frac{1}{12} a_{M_q}\right),$$

$$\theta^2_{\tilde{a}\tilde{A}} = b\left(a + (-1)^p \frac{1}{6} a_{M_p}\right) + b_r\left(\frac{1}{6}a + (-1)^q \frac{1}{12} a_{M_p}\right),$$

$$q = \begin{cases} 1, b - b_l \geq 0 \\ 2, b + b_r < 0 \end{cases}, M_q = \begin{cases} l, q = 1 \\ r, q = 2 \end{cases}, p = \begin{cases} 2, b - b_l \geq 0 \\ 1, b + b_r < 0 \end{cases}, M_p = \begin{cases} l, p = 1 \\ r, p = 2 \end{cases}.$$

Let determine aggregation intervals $\left[A_1^L, A_2^L\right], \left[A_1^U, A_2^U\right]$ for low membership function $\underline{\mu_{\tilde{A}}} = \left(a^L, a_l^L, a_r^L\right)$ and upper membership function $\overline{\mu_{\tilde{A}}} = \left(a^U, a_l^U, a_r^U\right)$ of $\tilde{A}$:

$$A_1^L = a^L - \frac{1}{6}a_l^L, A_2^L = a^L + \frac{1}{6}a_r^L,$$

$$A_1^U = a^U - \frac{1}{6}a_l^U, A_2^U = a^U + \frac{1}{6}a_r^U.$$

Let $f^2\left(\tilde{A}, \tilde{B}\right) = \left(A_1^L - B_1^L\right)^2 + \left(A_2^L - B_2^L\right)^2 + \left(A_1^U - B_1^U\right)^2 + \left(A_2^U - B_2^U\right)^2$, where $\left[A_1^L, A_2^L\right], \left[A_1^U, A_2^U\right], \left[B_1^L, B_2^L\right], \left[B_1^U, B_2^U\right]$ are weighed intervals for output $\tilde{Y}_i i = \overline{1, n}$, defined by low membership functions $\underline{\mu_{\tilde{Y}_i}} = \left(y^{iL}, y_l^{iL}, y_r^{iL}\right)$, $i = \overline{1, n}$ and upper membership functions $\overline{\mu_{\tilde{Y}_i}} = \left(y^{iU}, y_l^{iU}, \overline{y_r^{iU}}\right)$, $y^{iU} - y_l^{iU} \geq 0\ i = \overline{1, n}$.

Let $\tilde{X}_j^i$, $j = \overline{1, m}$, $i = \overline{1, n}$ input interval type-II fuzzy sets, defined by low membership functions $\underline{\mu_{\tilde{X}_j^i}} = \left(x^{jiL}, x_l^{jiL}, x_r^{jiL}\right)$ and upper membership functions $\overline{\mu_{\tilde{X}_j^i}} = \left(x^{jiU}, x_l^{jiU}, \overline{x_r^{jiU}}\right)$, $x^{jiU} - x_l^{jiU} \geq 0$, $j = \overline{1, m}, i = \overline{1, n}$.

We will construct a fuzzy regression model as follows:

$$\tilde{Y} = \tilde{a}_0 + \tilde{a}_1 \tilde{X}_1 + \cdots + \tilde{a}_m \tilde{X}_m,$$

where $\tilde{a}_j \equiv \left(b^j, b_l^j, b_r^j\right)$, $j = \overline{0, m}$—are triangular fuzzy numbers.

Let us determine the weighed intervals $\left[\theta^{1L}_{\hat{Y}_i}, \theta^{2L}_{\hat{Y}_i}\right], \left[\theta^{1U}_{\hat{Y}_i}, \theta^{2U}_{\hat{Y}_i}\right]$ for low membership functions and upper membership functions of $\hat{Y}_i = \tilde{a}_0 + \tilde{a}_1 \tilde{X}_1^i + \cdots + \tilde{a}_m \tilde{X}_m^i$:

$$\theta^{1L}_{\hat{Y}_i} = b^0 - \frac{1}{6}b_l^0 + \sum_{j=1}^{m} \theta^{1L}_{\tilde{a}_j \tilde{X}_j^i}\left(b^j, b_l^j, b_r^j\right),$$

$$\theta_{\hat{Y}_i}^{2L} = b^0 - \frac{1}{6}b_l^0 + \sum_{j=1}^{m} \theta_{\tilde{a}_j \tilde{X}_j^i}^{2L}\left(b^j, b_l^j, b_r^j\right),$$

$$\theta_{\hat{Y}_i}^{1U} = b^0 - \frac{1}{6}b_l^0 + \sum_{j=1}^{m} \theta_{\tilde{a}_j \tilde{X}_j^i}^{1U}\left(b^j, b_l^j, b_r^j\right),$$

$$\theta_{\hat{Y}_i}^{2U} = b^0 - \frac{1}{6}b_l^0 + \sum_{j=1}^{m} \theta_{\tilde{a}_j \tilde{X}_j^i}^{2U}\left(b^j, b_l^j, b_r^j\right),$$

$$\theta_{\tilde{a}_j \tilde{X}_j^i}^{1L}\left(b^j, b_l^j, b_r^j\right) = b^j\left(x^{jiL} + (-1)^q \frac{1}{6} x_{M_q}^{jiL}\right) - b_l^j\left(\frac{1}{6}x^{jiL} + (-1)^q \frac{1}{12} x_{M_q}^{jiL}\right),$$

$$\theta_{\tilde{a}_j \tilde{X}_j^i}^{2L}\left(b^j, b_l^j, b_r^j\right) = b^j\left(x^{jiL} + (-1)^p \frac{1}{6} x_{M_p}^{jiL}\right) + b_r^j\left(\frac{1}{6}x^{jiL} + (-1)^q \frac{1}{12} x_{M_p}^{jiL}\right),$$

$$\theta_{\tilde{a}_j \tilde{X}_j^i}^{1U}\left(b^j, b_l^j, b_r^j\right) = b^j\left(x^{jiU} + (-1)^q \frac{1}{6} x_{M_q}^{jiU}\right) - b_l^j\left(\frac{1}{6}x^{jiU} + (-1)^q \frac{1}{12} x_{M_q}^{jiU}\right),$$

$$\theta_{\tilde{a}_j \tilde{X}_j^i}^{2U}\left(b^j, b_l^j, b_r^j\right) = b^j\left(x^{jiU} + (-1)^p \frac{1}{6} x_{M_p}^{jiU}\right) + b_r^j\left(\frac{1}{6}x^{jiU} + (-1)^q \frac{1}{12} x_{M_p}^{jiU}\right),$$

$$q = \begin{cases} 1, b - b_l \geq 0 \\ 2, b + b_r < 0 \end{cases}, M_q = \begin{cases} l, q = 1 \\ r, q = 2 \end{cases}, p = \begin{cases} 2, b - b_l \geq 0 \\ 1, b + b_r < 0 \end{cases}, M_p = \begin{cases} l, p = 1 \\ r, p = 2 \end{cases}.$$

It is complicated enough to determine low membership functions and upper membership functions of model output data because it is not always triangular fuzzy numbers. That is why we will use the definition of $\alpha$-cuts for a fuzzy number.

If $\tilde{a} \equiv \left(b, b_l, b_r\right) < 0$, $\tilde{A} = \left(a_1, a_{2,} a_l, a_r\right) \geq 0$ then $\alpha$-cut $\left[C_\alpha^1, C_\alpha^2\right]$ of $\tilde{a}\tilde{A}$ looks like:

$$C_\alpha^1 = ba + (1-\alpha)ba_r - (1-\alpha)b_l a - (1-\alpha)^2 b_l a_r,$$

$$C_\alpha^2 = ba + (1-\alpha)ba_l + (1-\alpha)b_r a - (1-\alpha)^2 b_r a_l.$$

If $\tilde{a} \equiv \left(b, b_l, b_r\right) \geq 0$, $\tilde{A} = \left(a_1, a_{2,} a_l, a_r\right) \geq 0$ then $\alpha$-cut $\left[C_\alpha^1, C_\alpha^2\right]$ of $\tilde{a}\tilde{A}$ looks like:

$$C_\alpha^1 = ba - (1-\alpha)ba_l - (1-\alpha)b_r a + (1-\alpha)^2 b_r a_l,$$

$$C_\alpha^2 = ba + (1-\alpha)ba_r + (1-\alpha)b_l a + (1-\alpha)^2 b_l a_r.$$

According to the definition of weighed interval we get weighed intervals $\left[\theta_{\tilde{Y}_i}^{1L}, \theta_{\tilde{Y}_i}^{2L}\right]$, $\left[\theta_{\tilde{Y}_i}^{1U}, \theta_{\tilde{Y}_i}^{2U}\right]$ for initial output data $\tilde{Y}_i i = \overline{1, n}$:

$$\theta_{\tilde{Y}_i}^{1L} = y^{iL} - \frac{1}{6}y_l^{iL}, \theta_{\tilde{Y}_i}^{2L} = y^{iL} + \frac{1}{6}y_r^{iL},$$

$$\theta_{\tilde{Y}_i}^{1U} = y^{iU} - \frac{1}{6} y_l^{iU}, \theta_{\tilde{Y}_i}^{2U} = y^{iU} + \frac{1}{6} y_r^{iU}.$$

We determine a functional

$$F\left(b^j, b_l^j, b_r^j\right) = \sum_{i=1}^{n} f^2\left(\hat{Y}_i, \tilde{Y}_i\right) = \sum_{i=1}^{n}\left[\left(\theta_{\hat{Y}_i}^{1L} - \theta_{\tilde{Y}_i}^{1L}\right)^2 + \left(\theta_{\hat{Y}_i}^{2L} - \theta_{\tilde{Y}_i}^{2L}\right)^2\right]$$
$$+ \sum_{i=1}^{n}\left[\left(\theta_{\hat{Y}_i}^{1U} - \theta_{\tilde{Y}_i}^{1U}\right)^2 + \left(\theta_{\hat{Y}_i}^{2U} - \theta_{\tilde{Y}_i}^{2U}\right)^2\right],$$

then

$$F\left(b^j, b_l^j, b_r^j\right) = \sum_{i=1}^{n}\left[b^0 - \frac{1}{6} b_l^0 - y^{iL} + \frac{1}{6} y_l^{iL} + \sum_{j=1}^{m} \theta_{\tilde{a}_j \tilde{X}_j^i}^{1L}\left(b^j, b_l^j, b_r^j\right)\right]^2 +$$
$$+ \sum_{i=1}^{n}\left[b^0 + \frac{1}{6} b_r^0 - y^{iL} - \frac{1}{6} y_r^{iL} + \sum_{j=1}^{m} \theta_{\tilde{a}_j \tilde{X}_j^i}^{2L}\left(b^j, b_L^j, b_R^j\right)\right]^2 +$$
$$+ \sum_{i=1}^{n}\left[b^0 - \frac{1}{6} b_l^0 - y^{iU} + \frac{1}{6} y_l^{iU} + \sum_{j=1}^{m} \theta_{\tilde{a}_j \tilde{X}_j^i}^{1U}\left(b^j, b_l^j, b_r^j\right)\right]^2 +$$
$$+ \sum_{i=1}^{n}\left[b^0 + \frac{1}{6} b_r^0 - y^{iU} - \frac{1}{6} y_r^{iU} + \sum_{j=1}^{m} \theta_{\tilde{a}_j \tilde{X}_j^i}^{2U}\left(b^j, b_l^j, b_r^j\right)\right]^2.$$

It is easy to see that $F\left(b^j, b_l^j, b_r^j\right)$ is piecewise differentiable function in the field $b_l^j \geq 0, b_r^j \geq 0, j = \overline{0, m}$ because $\theta_{\tilde{a}_j \tilde{X}_j^i}^{1L}\left(b^j, b_l^j, b_r^j\right)$, $\theta_{\tilde{a}_j \tilde{X}_j^i}^{2L}\left(b^j, b_l^j, b_r^j\right)$, $\theta_{\tilde{a}_j \tilde{X}_j^i}^{1U}\left(b^j, b_l^j, b_r^j\right)$, $\theta_{\tilde{a}_j \tilde{X}_j^i}^{2U}\left(b^j, b_l^j, b_r^j\right)$ are piecewise linear functions.

We will find unknown parameters from the condition:

$$F\left(b^j, b_l^j, b_r^j\right) = \sum_{i=1}^{n} f^2\left(\hat{Y}_i, \tilde{Y}_i\right) \to \min,$$
$$b_l^j \geq 0, b_r^j \geq 0, j = \overline{0, m}$$

by known methods [30].

As quality indicators of the regression model play a significant role, we define by analogy with the classical regression model the standard deviation of the output variable ($S_{\tilde{y}}$), the correlation coefficient ($HR^2$) and the standard error of estimates of the output variable:

$$S_{\tilde{y}} = \sqrt{\frac{1}{n-1}\sum_{i=1}^{n} f^2\left(\tilde{Y}_i, \bar{\tilde{Y}}\right)}, \bar{\tilde{Y}} = \frac{\sum_{i=1}^{n} \tilde{Y}_i}{n},$$

$$HR^2 = \frac{\sum_{i=1}^{n} f^2\left(\hat{Y}_i, \bar{\tilde{Y}}\right)}{\sum_{i=1}^{n} f^2\left(\tilde{Y}_i, \bar{\tilde{Y}}\right)},$$

$$HS = \sqrt{\frac{1}{n-m-1}\sum_{i=1}^{n} f^2\left(\hat{Y}_i, \tilde{Y}_i\right)}.$$

Suppose that for evaluation output parameter $Y$ experts use a linguistic scale with levels $Y_k$, $k = \overline{1, p}$, that are formalized with the help of interval type-II fuzzy sets $\tilde{\tilde{Y}}_k$, $k = \overline{1, p}$ defined by their low membership functions $\mu_{\tilde{\tilde{Y}}_k} = \left(y^{kL}, y_l^{kL}, y_r^{kL}\right)$, $k = \overline{1, p}$ and upper membership functions $\overline{\mu_{\tilde{\tilde{Y}}_k}} = \left(y^{kU}, y_l^{kU}, y_r^{kU}\right)$, $k = \overline{1, p}$. We have got model output value $\hat{Y}_i$ of regression model in the form of interval type-II fuzzy set, defined by low membership function $\mu_{\hat{Y}_i} = \left(v^{iL}, v_l^{iL}, v_r^{iL}\right)$ and upper membership function $\overline{\mu_{\hat{Y}_i}} = \left(v^{iU}, v_l^{iU}, v_r^{iU}\right)$. But it is very important to identify this fuzzy set with one of the levels $Y_k$, $k = \overline{1, p}$ of the linguistic scale used by experts.

Let $\left[C_1^{iL}, C_2^{iL}\right]$, $\left[C_1^{iL}, C_2^{iL}\right]$, $i = \overline{1, n}$ are weighted intervals of $\hat{Y}_i$, $\left[D_1^{iL}, D_2^{iL}\right]$, $\left[D_1^{iL}, D_2^{iL}\right]$, $k = \overline{1, p}$ are weighted intervals of $\tilde{\tilde{Y}}_k$, $k = \overline{1, p}$.

Then

$$f^2\left(\hat{Y}_i, \tilde{\tilde{Y}}_k\right) = \left(C_1^{iL} - D_1^{kL}\right)^2 + \left(C_2^{iL} - D_2^{kL}\right)^2 + \\ + \left(C_1^{iU} - D_1^{iU}\right)^2 + \left(C_2^{iU} - D_2^{iU}\right)^2, i = \overline{1, n}, k = \overline{1, p}.$$

$\hat{Y}_i$ is identified to level $Y_s$ of the linguistic scale used, if

$$f^2\left(\hat{Y}_i, \tilde{\tilde{Y}}_s\right) = \min_k f^2\left(\hat{Y}_i, \tilde{\tilde{Y}}_k\right), k = \overline{1, p}.$$

## 7 Prediction of Expert Evaluations Based on Nonlinear Regression with Initial Interval Type-II Data

Let $\tilde{Y}_i i = \overline{1, n}$ are output interval type-II fuzzy sets, defined by low membership functions $\mu_{\tilde{Y}_i} = \left(y_1^{iL}, y_2^{iL}, y_l^{iL}, y_r^{iL}\right)$, $i = \overline{1, n}$ and upper membership functions $\overline{\mu_{\tilde{Y}_i}} = \left(y_1^{iU}, y_2^{iU}, y_l^{iU}, y_r^{iU}\right)$, $y_1^{iU} - y_l^{iU} \geq 0 i = \overline{1, n}$.

Let $\tilde{X}_i$, $i = \overline{1,n}$ input interval type-II fuzzy sets, defined by low membership functions $\underline{\mu_{\tilde{X}_i}} = \left(x_1^{iL}, x_2^{iL}, x_l^{iL}, x_r^{iL}\right)$, $i = \overline{1,n}$ and upper membership functions $\overline{\mu_{\tilde{X}_i}} = \left(x_1^{iU}, x_2^{iU}, x_l^{iU}, x_r^{iU}\right)$, $x_1^{iU} - x_l^{iU} \geq 0, i = \overline{1,n}$.

We will construct a fuzzy regression model as follows:

$$\tilde{Y} = \tilde{a}_0 + \tilde{a}_1\tilde{X} + \tilde{a}_2\tilde{X}^2,$$

where $\tilde{a}_j \equiv \left(b^j, b_l^j, b_r^j\right)$, $j = \overline{0,2}$—are triangular fuzzy numbers.

We determine the weighed intervals $\left[\theta_{\hat{Y}_i}^{1L}, \theta_{\hat{Y}_i}^{2L}\right]$, $\left[\theta_{\hat{Y}_i}^{1U}, \theta_{\hat{Y}_i}^{2U}\right]$ for low membership functions and upper membership functions of $\tilde{Y} = \tilde{a}_0 + \tilde{a}_1\tilde{X} + \tilde{a}_2\tilde{X}^2$:

$$\theta_{\hat{Y}_i}^{1L} = b^0 - \frac{1}{6}b_l^0 + \theta_{\tilde{a}_1\tilde{X}_i}^{1L}\left(b^1, b_l^1, b_r^1\right) + \theta_{\tilde{a}_2\tilde{X}_i^2}^{1L}\left(b^2, b_l^2, b_r^2\right),$$

$$\theta_{\hat{Y}_i}^{2L} = b^0 - \frac{1}{6}b_l^0 + \theta_{\tilde{a}_1\tilde{X}_i}^{2L}\left(b^1, b_l^1, b_r^1\right) + \theta_{\tilde{a}_2\tilde{X}_i^2}^{2L}\left(b^2, b_l^2, b_r^2\right),$$

$$\theta_{\hat{Y}_i}^{1U} = b^0 - \frac{1}{6}b_l^0 + \theta_{\tilde{a}_1\tilde{X}_i}^{1U}\left(b^1, b_l^1, b_r^1\right) + \theta_{\tilde{a}_2\tilde{X}_i^2}^{1U}\left(b^2, b_l^2, b_r^2\right),$$

$$\theta_{\hat{Y}_i}^{2U} = b^0 - \frac{1}{6}b_l^0 + \theta_{\tilde{a}_1\tilde{X}_i}^{2U}\left(b^1, b_l^1, b_r^1\right) + \theta_{\tilde{a}_2\tilde{X}_i^2}^{2U}\left(b^2, b_l^2, b_r^2\right).$$

We determine the weighed intervals $\left[\theta_{\tilde{Y}_i}^{1L}, \theta_{\tilde{Y}_i}^{2L}\right]$, $\left[\theta_{\tilde{Y}_i}^{1U}, \theta_{\tilde{Y}_i}^{2U}\right]$ for low membership functions and upper membership functions of $\tilde{Y}_i i = \overline{1,n}$:

$$\theta_{\tilde{Y}_i}^{1L} = y_1^{iL} - \frac{1}{6}y_l^{iL}, \theta_{\tilde{Y}_i}^{2L} = y_2^{iL} + \frac{1}{6}y_r^{iL},$$

$$\theta_{\tilde{Y}_i}^{1U} = y_1^{iU} - \frac{1}{6}y_l^{iU}, \theta_{\tilde{Y}_i}^{2U} = y_2^{iU} + \frac{1}{6}y_r^{iU}.$$

We determine a functional

$$F\left(b^j, b_l^j, b_r^j\right) = \sum_{i=1}^{n} f^2\left(\hat{Y}_i, \tilde{Y}_i\right) = $$

$$= \sum_{i=1}^{n}\left[\left(\theta_{\hat{Y}_i}^{1L} - \theta_{\tilde{Y}_i}^{1L}\right)^2 + \left(\theta_{\hat{Y}_i}^{2L} - \theta_{\tilde{Y}_i}^{2L}\right)^2\right] +$$

$$+ \sum_{i=1}^{n}\left[\left(\theta_{\hat{Y}_i}^{1U} - \theta_{\tilde{Y}_i}^{1U}\right)^2 + \left(\theta_{\hat{Y}_i}^{2U} - \theta_{\tilde{Y}_i}^{2U}\right)^2\right].$$

It is easy to see that $F\left(b^j, b_l^j, b_r^j\right)$ is piecewise differentiable function in the field $b_l^j \geq 0$, $b_r^j \geq 0$, $j = \overline{1,2}$ because $\theta_{\tilde{a}_1\tilde{X}_i}^{1L}\left(b^1, b_l^1, b_r^1\right)$, $\theta_{\tilde{a}_1\tilde{X}_i}^{2L}\left(b^1, b_l^1, b_r^1\right)$,

$\theta^{1U}_{\tilde{a}_1\tilde{X}_i}(b^1, b^1_l, b^1_r)$, $\theta^{2U}_{\tilde{a}_1\tilde{X}_i}(b^1, b^1_l, b^1_r)$, $\theta^{1L}_{\tilde{a}_2\tilde{X}_i^2}(b^2, b^2_l, b^2_r)$, $\theta^{2L}_{\tilde{a}_2\tilde{X}_i^2}(b^2, b^2_l, b^2_r)$, $\theta^{1U}_{\tilde{a}_2\tilde{X}_i^2}(b^2, b^2_l, b^2_r)$ $\theta^{2U}_{\tilde{a}_2\tilde{X}_i^2}(b^2, b^2_l, b^2_r)$ are piecewise linear functions.

We will find unknown parameters from the condition:

$$F\left(b^j, b^j_l, b^j_r\right) = \sum_{i=1}^{n} f^2\left(\hat{Y}_i, \tilde{Y}_i\right) \to \min,$$

$$b^j_l \geq 0, b^j_r \geq 0, j = \overline{0, 2}$$

by known methods [30].

As quality indicators of the regression model play a significant role, we define by analogy with the classical regression model the standard deviation of the output variable ($S_{\tilde{y}}$), the correlation coefficient ($HR^2$) and the standard error of estimates of the output variable:

$$S_{\tilde{y}} = \sqrt{\frac{1}{n-1}\sum_{i=1}^{n} f^2\left(\tilde{Y}_i, \bar{\tilde{Y}}\right)}, \bar{\tilde{Y}} = \frac{\sum_{i=1}^{n}\tilde{Y}_i}{n},$$

$$HR^2 = \frac{\sum_{i=1}^{n} f^2\left(\hat{Y}_i, \bar{\tilde{Y}}\right)}{\sum_{i=1}^{n} f^2\left(\tilde{Y}_i, \bar{\tilde{Y}}\right)}, HS = \sqrt{\frac{1}{n-2}\sum_{i=1}^{n} f^2\left(\hat{Y}_i, \tilde{Y}_i\right)}.$$

Suppose that for evaluation output parameter $Y$ experts use a linguistic scale with levels $Y_k$, $k = \overline{1, p}$, that are formalized with the help of interval type-II fuzzy sets $\tilde{\tilde{Y}}_k$, $k = \overline{1, p}$ defined by their low membership functions $\underline{\mu}_{\tilde{\tilde{Y}}_k} = \left(y_1^{kL}, y_2^{kL}, y_l^{kL}, y_r^{kL}\right)$, $k = \overline{1, p}$ and upper membership functions $\overline{\mu_{\tilde{\tilde{Y}}_k}} = \left(y_1^{kU}, y_2^{kU}, y_l^{kU}, y_r^{kU}\right)$, $k = \overline{1, p}$. We have got model output value $\hat{Y}_i$ of regression model in the form of interval type-II fuzzy set, defined by low membership function $\underline{\mu}_{\hat{Y}_i} = \left(v_1^{iL}, v_2^{iL}, v_l^{iL}, v_r^{iL}\right)$ and upper membership function $\overline{\mu_{\hat{Y}_i}} = \left(v_1^{iU}, v_2^{iU}, v_l^{iU}, v_r^{iU}\right)$. But it is very important to identify this fuzzy set with one of the levels $Y_k$, $k = \overline{1, p}$ of the linguistic scale used by experts.

Let $\left[C_1^{iL}, C_2^{iL}\right]$, $\left[C_1^{iL}, C_2^{iL}\right]$, $i = \overline{1, n}$ are weighted intervals of $\hat{Y}_i$, $\left[D_1^{iL}, D_2^{iL}\right]$, $\left[D_1^{iL}, D_2^{iL}\right]$, $k = \overline{1, p}$ are weighted intervals of $\tilde{\tilde{Y}}_k$, $k = \overline{1, p}$.

Then

$$f^2\left(\hat{Y}_i, \tilde{\tilde{Y}}_k\right) = \left(C_1^{iL} - D_1^{kL}\right)^2 + \left(C_2^{iL} - D_2^{kL}\right)^2 + \left(C_1^{iU} - D_1^{iU}\right)^2 + \left(C_2^{iU} - D_2^{iU}\right)^2, i = \overline{1, n}, k = \overline{1, p}.$$

The value of regression model $\hat{Y}_i$ is identified to level $Y_s$ of the linguistic scale used, if

$$f^2\left(\hat{Y}_i, \tilde{\tilde{Y}}_s\right) = \min_k f^2\left(\hat{Y}_i, \tilde{\tilde{Y}}_k\right), k = \overline{1, p}.$$

## 8 Prediction of Quantitative Parameters Values Based on Linear Regression with Interval Type-II Coefficients

Let $y_i$, $i = \overline{1, n}$ are output crisp data—the values of a certain numerical parameter $Y$, $x_j^i$, $j = \overline{1, m}$, $i = \overline{1, n}$ input crisp data -the values of some numerical parameters $X_1, \ldots, X_m$.

We will construct a fuzzy regression model as follows [35]:

$$Y = \tilde{a}_0 + \tilde{a}_1 X_1 + \cdots + \tilde{a}_m X_m.$$

$\tilde{a}_j$, $j = \overline{0, m}$—interval type-II fuzzy sets, defined by $\underline{\tilde{a}_j} \equiv \left(a_1^{jL}, a_2^{jL}, a_l^{jL}, a_r^{jL}\right)$, $j = \overline{0, m}$ and $\overline{\tilde{a}_j} \equiv \left(a_1^{jU}, a_2^{jU}, a_l^{jU}, a_r^{jU}\right)$, $j = \overline{0, m}$.

The weighed interval $\left[\theta^1_{x\tilde{A}}, \theta^2_{x\tilde{A}}\right]$ of product of crisp number $x \geq 0$ and trapezoidal fuzzy number $\tilde{A} \equiv (a_1, a_2, a_l, a_r)$ looks like

$$\theta^1_{x\tilde{A}} = x\left(a_q + (-1)^q \frac{1}{6} a_{M_q}\right), \theta^2_{x\tilde{A}} = x\left(a_p + (-1)^p \frac{1}{6} a_{M_p}\right)$$

$$q = \begin{cases} 1, a_1 - a_l \geq 0 \\ 2, a_2 + a_r < 0 \end{cases}, M_q = \begin{cases} l, q = 1 \\ r, q = 2 \end{cases}, p = \begin{cases} 2, a_1 - a_l \geq 0 \\ 1, a_2 + a_r < 0 \end{cases}, M_p = \begin{cases} l, p = 1 \\ r, p = 2 \end{cases}.$$

The weighed interval $\left[\theta^1_{x\tilde{A}}, \theta^2_{x\tilde{A}}\right]$ of product of crisp number $x \leq 0$ and trapezoidal fuzzy number $\tilde{A} \equiv (a_1, a_2, a_l, a_r)$ looks like

$$\theta^1_{x\tilde{A}} = x\left(a_p + (-1)^p \frac{1}{6} a_{M_p}\right), \theta^2_{x\tilde{A}} = x\left(a_q + (-1)^q \frac{1}{6} a_{M_q}\right)$$

$$q = \begin{cases} 1, a_1 - a_l \geq 0 \\ 2, a_2 + a_r < 0 \end{cases}, M_q = \begin{cases} l, q = 1 \\ r, q = 2 \end{cases}, p = \begin{cases} 2, a_1 - a_l \geq 0 \\ 1, a_2 + a_r < 0 \end{cases}, M_p = \begin{cases} l, p = 1 \\ r, p = 2 \end{cases}.$$

We determine the weighed intervals $\left[\theta^{1L}_{\hat{Y}_i}, \theta^{2L}_{\hat{Y}_i}\right], \left[\theta^{1U}_{\hat{Y}_i}, \theta^{2U}_{\hat{Y}_i}\right]$ for $\tilde{Y}_i = \tilde{a}_0 + \tilde{a}_1 X_1^i + \cdots + \tilde{a}_m X_m^i$:

$$\theta^{1L}_{\hat{Y}_i} = a_1^0 - \frac{1}{6} a_l^0 + \sum_{j=1}^{m} \theta^{1L}_{\tilde{a}_j X_j^i}\left(a_1^{jL}, a_2^{jL}, a_l^{jL}, a_r^{jL}\right),$$

$$\theta_{\hat{Y}_i}^{2L} = a_2^0 + \frac{1}{6}a_r^0 + \sum_{j=1}^{m} \theta_{\tilde{a}_j X_j^i}^{2L}\left(a_1^{jL}, a_2^{jL}, a_l^{jL}, a_r^{jL}\right),$$

$$\theta_{\hat{Y}_i}^{1U} = a_1^0 - \frac{1}{6}a_l^0 + \sum_{j=1}^{m} \theta_{\tilde{a}_j X_j^i}^{1U}\left(a_1^{jU}, a_2^{jU}, a_l^{jU}, a_r^{jU}\right),$$

$$\theta_{\hat{Y}_i}^{2U} = a_2^0 + \frac{1}{6}a_r^0 + \sum_{j=1}^{m} \theta_{\tilde{a}_j X_j^i}^{2U}\left(a_1^{jU}, a_2^{jU}, a_l^{jU}, a_r^{jU}\right).$$

We determine a functional

$$F\left(a_1^{jL}, a_2^{jL}, a_l^{jL}, a_r^{jL}, a_1^{jU}, a_2^{jU}, a_l^{jU}, a_r^{jU}\right) = \sum_{i=1}^{n} f^2\left(y_i, \tilde{Y}_i\right) =$$

$$= \sum_{i=1}^{n}\left[\left(y_i - \theta_{\tilde{Y}_i}^{1L}\right)^2 + \left(y_i - \theta_{\tilde{Y}_i}^{2L}\right)^2\right] +$$

$$+ \sum_{i=1}^{n}\left[\left(y_i - \theta_{\tilde{Y}_i}^{1U}\right)^2 + \left(y_i - \theta_{\tilde{Y}_i}^{2U}\right)^2\right],$$

It is easy to see that $F\left(a_1^{jL}, a_2^{jL}, a_l^{jL}, a_r^{jL}, a_1^{jU}, a_2^{jU}, a_l^{jU}, a_r^{jU}\right)$ is piecewise differentiable function in the field $a_l^{jL} \geq 0$, $a_r^{jL} \geq 0$, $a_l^{jU} \geq 0$, $a_r^{jU} \geq 0$, $j = \overline{0, m}$ because $\theta_{\tilde{a}_1 \tilde{X}_i}^{1L}(b^1, b_l^1, b_r^1)$, $\theta_{\tilde{a}_1 \tilde{X}_i}^{2L}(b^1, b_l^1, b_r^1)$, $\theta_{\tilde{a}_1 \tilde{X}_i}^{1U}(b^1, b_l^1, b_r^1)$, $\theta_{\tilde{a}_1 \tilde{X}_i}^{2U}(b^1, b_l^1, b_r^1)$, $\theta_{\tilde{a}_2 \tilde{X}_i^2}^{1L}(b^2, b_l^2, b_r^2)$, $\theta_{\tilde{a}_2 \tilde{X}_i^2}^{2L}(b^2, b_l^2, b_r^2)$, $\theta_{\tilde{a}_2 \tilde{X}_i^2}^{1U}(b^2, b_l^2, b_r^2)$ $\theta_{\tilde{a}_2 \tilde{X}_i^2}^{2U}(b^2, b_l^2, b_r^2)$ are piecewise linear functions.

We will find unknown parameters from the condition:

$$F\left(a_1^{jL}, a_2^{jL}, a_l^{jL}, a_r^{jL}, a_1^{jU}, a_2^{jU}, a_l^{jU}, a_r^{jU}\right) = \sum_{i=1}^{n} f^2\left(y_i, \tilde{Y}_i\right) \to \min,$$

$$a_l^{jL} \geq 0, a_r^{jL} \geq 0, a_l^{jU} \geq 0, a_r^{jU} \geq 0, j = \overline{0, m}.$$

by known methods [30].

Let $\hat{Y}_i i = \overline{1, n}$ are model output interval type-II fuzzy sets, defined by low membership functions $\underline{\mu_{\hat{Y}_i}} = \left(v_1^{iL}, v_2^{iL}, v_l^{iL}, v_r^{iL}\right), i = \overline{1, n}$ and upper membership functions $\overline{\mu_{\hat{Y}_i}} = \left(v_1^{iU}, v_2^{iU}, v_l^{iU}, v_r^{iU}\right), i = \overline{1, n}$. After obtaining $\hat{Y}_i i = \overline{1, n}$ a problem of identifying them appears.

The weighted intervals for low membership function and upper membership function of model $\hat{Y}_i$, $i = \overline{1, n}$ are designated by $\left[C_1^{iL}, C_2^{iL}\right]$, $\left[C_1^{iU}, C_2^{iU}\right]$, $i = \overline{1, n}$ accordingly.

$$C_1^{iL} = v_1^{iL} - \frac{1}{6}v_l^{iL}, C_2^{iL} = v_2^{iL} + \frac{1}{6}v_r^{iL},$$

$$C_1^{iU} = \nu_1^{iU} - \frac{1}{6}\nu_l^{iU}, C_2^{iU} = \nu_2^{iU} + \frac{1}{6}\nu_r^{iU}.$$

$\hat{Y}_i i = \overline{1, n}$ is identified as $\frac{1}{4}\left(C_1^{iL} + C_2^{iL} + C_1^{iU} + C_2^{iU}\right)$.

As quality indicators of the regression model play a significant role, we define the standard deviation of the output variable ($S_{\bar{y}}$), the correlation coefficient ($HR^2$) and the standard error of estimates of the output variable:

$$S_y = \sqrt{\frac{1}{n-1}\sum_{i=1}^{n}(y_i - \overline{y})^2}, \overline{y} = \frac{\sum_{i=1}^{n} y_i}{n},$$

$$HR^2 = \frac{\sum_{i=1}^{n} f^2\left(\hat{Y}_i, \overline{y}\right)}{\sum_{i=1}^{n}(y_i - \overline{y})^2},$$

$$HS = \sqrt{\frac{1}{n-m-1}\sum_{i=1}^{n} f^2\left(\hat{Y}_i, y_i\right)}.$$

## 9 Conclusions

The purpose of this chapter is to propose the methods of formalization, prediction and recognition which help experts to find the anomaly symptoms by extracting important information from the telemetry data. The participation of experts is extremely important when evaluating complex technical objects under conditions of heterogeneous uncertainty. Experts often use verbal scales to assess quantitative and qualitative parameters. The values or levels of these scales are the words of the professional language of experts.

A method for formalizing group expert information on the basis of interval type-II fuzzy sets is developed, which significantly complement the methods of formalization of expert information developed on the basis of type-I fuzzy sets. This method allows you to obtain not averaged expert opinion, but take into account all the essential information received from each expert. For the prediction of expert evaluations, regression models based on interval type-II fuzzy sets were developed. The first model is linear and allows predicting expert evaluations of qualitative parameters. The second model is developed for a special class interval type-II fuzzy sets, which can simplify the procedures of expert evaluation. The third model is nonlinear and allows predicting expert evaluations of qualitative parameters. The fourth model with interval type-II fuzzy coefficients is developed for prediction numerical parameters. Construction of all the models based on definition of weighted intervals for input and output data. As quality indicators of the regression model play a significant role, we define by analogy with the classical regression model the standard deviation of the output variable, the correlation coefficient and the standard error of estimates of the output variable.

## References


1. L.A. Zadeh, Fuzzy sets. Inf. Control **8**, 338–352 (1965)
2. L.A. Zadeh, The Concept of a linguistic variable and its application to approximate reasoning, part 1, 2 and 3. Inf. Sci. **8**, 199–249, 301–357 (1975); Inf. Sci. **9**, 43–80 (1976)
3. A.N. Averkin, I.Z. Batyrshin, A.F. Blishun, V.B. Silov, V.B. Tarasov, *Fuzzy Sets in Models of Control and Artificial Intelligence* (Main Office on Physical-Mathematical Literature, Nauka, Moscow, 1986) (in Russian)
4. A.N. Borisov, O.A. Krumberg, I.P. Fedorov, *Decision Making on the Basis of Fuzzy Models: Examples of Use* (Zinatne, Riga, 1990), 184 pp.
5. O. Poleshchuk, E. Komarov, *Expert Fuzzy Information Processing*. Studies in Fuzziness and Soft Computing, vol. 268 (2011), pp. 1–239
6. A.E. Altunin, M.V. Semukhin, *Models and Algorithms of a Decision Making in Fuzzy Conditions* (Publishing house of Tyumen State University, Tyumen, 2002), 268 pp. (in Russian)
7. A.P. Ryjov, The concept of a full orthogonal semantic scope and the measuring of semantic uncertainty, in *Fifth International Conference Information Processing and Management of Uncertainty in Knowledge-Based Systems* (1994), pp. 33–34
8. A. Ryjov, Fuzzy linguistic scales: definition, properties and applications, in *Soft Computing in Measurement and Information Acquisition*, ed. by L. Reznik, V. Kreinovich. Studies in Fuzziness and Soft Computing, vol. 127 (2003)
9. O. Poleshchuk, The determination of students' fuzzy rating points and qualification levels. Int. J. Ind. Syst. Eng. **9**(1), 3–20 (2011)
10. A. Darwish, O. Poleshchuk, New models for monitoring and clustering of the state of plant species based on sematic spaces. J. Intell. Fuzzy Syst. **26**(3), 1089–1094 (2014)
11. C.L. Hwang, N.J. Lin, *Group Decision Making Under Multiple Criteria* (Springer, Berlin, 1987), 400 pp.
12. F. Liu, J.M. Mendel, Encoding words into interval Type-2 fuzzy sets using an interval approach. IEEE Trans. Fuzzy Syst. **16**(6) (2008)
13. Y.-H.O. Chang, Hybrid fuzzy least-squares regression analysis and its reliability measures. Fuzzy Sets Syst. **119**, 225–246 (2001)
14. O.M. Poleshuk, E.G. Komarov, New defuzzification method based on weighted intervals, in *Annual Conference of the North American Fuzzy Information Processing Society, NAFIPS'2008* (2008), p. 14531223. https://doi.org/10.1109/nafips.2008.4531223
15. O. Poleshchuk, E. Komarov, A nonlinear hybrid fuzzy least-squares regression model, in *Annual Conference of the North American Fuzzy Information Processing Society—NAFIPS'2011*, El Paso, Texas, 18–20 Mar 2011. https://doi.org/10.1109/nafips.2011.5751909
16. H. Tanaka, Fuzzy data analysis by possibilistic linear models. Fuzzy Sets Syst. **21**, 363–375 (1991)
17. H. Tanaka, S. Uejima, K. Asai, Linear regression analysis with fuzzy model. IEEE. Trans. Syst. Man Cybern. **SMC-2**, 903–907 (1982)
18. H. Tanaka, H. Ishibuchi, Identification of possibilistic linear models. Fuzzy Sets Syst. **41**, 145–160 (1991)
19. H. Tanaka, H. Ishibuchi, S. Yoshikawa, Exponential possibility regression analysis. Fuzzy Sets Syst. **69**, 305–318 (1995)
20. A. Celmins, Least squares model fitting to fuzzy vector data. Fuzzy Sets Syst. **22**, 245–269 (1987)
21. A. Celmins, Multidimensional least-squares model fitting of fuzzy models. Math. Modeling **9**, 669–690 (1987)

22. D.A. Sabic, W. Pedrycr, Evaluation on fuzzy linear regression models. Fuzzy Sets Syst. **39**, 51–63 (1991)
23. Y.-H.O. Chang, Synthesize fuzzy-random data by hybrid fuzzy least-squares regression analysis. J. Natl. Kaohsiung Inst. Technol. **28**, 1–14 (1997)
24. Y.-H.O. Chang, Hybrid fuzzy-random analysis for system modeling. J. Natl. Kaohsiung Inst. Technol. **29**, 1–9 (1998)
25. Y.-H.O. Chang, B.M. Ayyub, Fuzzy regression methods—a comparative assessment. Fuzzy Sets Syst. **119**, 187–203 (2001)
26. R.J. Hathaway, J.C. Bezdek, Switching regression models and fuzzy clustering. IEEE Trans. Fuzzy Syst. **1**(3), 195–203 (1993)
27. I.B. Turksen, Fuzzy functions with LSE. Appl. Soft Comput. **8**(3), 1178–1188 (2008)
28. A. Celikyilmaz, Fuzzy functions with support vector machines. M.A.Sc. Thesis, Information Science, Industrial Engineering Department, University of Toronto (2005)
29. A. Celikyilmaz, I.B. Turksen, Fuzzy functions with support vector machines. Inf. Sci. **177**, 5163–5177 (2007)
30. T.F. Coleman, Y. Li, A reflective newton method for minimizing a quadratic function subject to bounds on some of the variables. SIAM J. Optim. **6**, 1040–1058 (1996)
31. O.M. Poleshuk, E.G. Komarov, Multiple hybrid regression for fuzzy observed data, in *Proceedings of the 27th International Conference of the North American Fuzzy Information Processing Society, NAFIPS'2008*, New York, New York, 19–22 May 2008
32. O. Poleshchuk, E. Komarov, *Hybrid Fuzzy Least-Squares Regression Model for Qualitative Characteristics*. Advances in Intelligent and Soft Computing, vol. 68 (2010), pp. 187–196
33. O. Poleshchuk, E. Komarov, A fuzzy linear regression model for interval type-2 fuzzy sets, in *Annual Conference of the North American Fuzzy Information Processing Society—NAFIPS'2012*. https://doi.org/10.1109/nafips.2012.6290970
34. A. Darwish, O. Poleshchuk, E. Komarov, A new fuzzy linear regression model for a special case of interval type-2 fuzzy sets. Appl. Math. Inf. Sci. **10**(3), 1209–1214 (2016). https://doi.org/10.18576/amis/100340
35. O.M. Poleshchuk, E.G. Komarov, A. Darwish, A fuzzy linear regression model with interval type-2 fuzzy coefficients, in *Proceedings of the 19th International Conference on Soft Computing and measurements (SCM)* (2016), pp. 388–391. https://doi.org/10.1109/scm.2016.7519789

# Intelligent Health Monitoring Systems for Space Missions Based on Data Mining Techniques

**Sara Abdelghafar, Ashraf Darwish and Aboul Ella Hassanien**

**Abstract** Development of intelligent health monitoring system for artificial satellites is one of the important issue of aerospace engineering, in which determines the health state and failure prediction of satellites based on the telemetry data. Recent development in data mining techniques make it possible to examine satellite telemetry data and extract embedded information to produce advanced system health monitoring applications. This study presents a framework of the essential operations and applications of intelligent health monitoring systems which are being applied in the ground control station. Furthermore, this study reviews an extensive collection of existing health monitoring solutions and discusses them in a framework of telemetry data mining techniques. The work presented in this study can be used as a guideline for designing intelligent health monitoring solution based on telemetry data mining techniques.

**Keywords** Satellite telemetry data mining · Health monitoring · Ground control operations

---

S. Abdelghafar (✉)
Computer Science Department, Faculty of Science, Al Azhar University, Cairo, Egypt
e-mail: sara.abdelghafar@yahoo.com
URL: http://www.egyptscience.net

A. Darwish
Faculty of Science, Helwan University, Cairo, Egypt
URL: http://www.egyptscience.net

A. E. Hassanien
IT Department, Faculty of Computers & Information, Cairo University, Giza, Egypt
URL: http://www.egyptscience.net

S. Abdelghafar · A. Darwish · A. E. Hassanien
Scientific Research Group in Egypt (SRGE), Giza, Egypt



A. E. Hassanien et al. (eds.), *Machine Learning and Data Mining in Aerospace Technology*, Studies in Computational Intelligence 836,
https://doi.org/10.1007/978-3-030-20212-5_4

## 1 Introduction

Satellites are amongst today's most complex technical systems, they fulfil their mission in a very special, harsh, and challenging environment. So it is practically impossible to completely eliminate the possibility of anomalies or faults. Also, determining the health state of these systems using conventional techniques based on expert knowledge or mode based diagnosis is becoming more difficult because of variety of factors external environment and performance degradation, like the status of satellite will change relative to the phase of design especially for long-life satellites. Therefore, telemetry data mining techniques have been developed to address the health monitoring operations. By analyzing system operations data to characterize and detect unusual behavior or anomalies, and analyzing the fault reason and the impact on the system. Then, followed by discovering the relationship between dependent and independent variables to predict the next event or behavior to ensure the highest level of safety and reliability. All previous processes will lead to the proper control of satellite based on the status of resources and mission operations. Several health monitoring approaches have been developed and successfully applied to aerospace operations for both real time system monitoring and archived data analysis. This study are focused on the main operations of health monitoring, detection, diagnosis and predication. In recent years, health monitoring operations based on satellite telemetry data mining have practical significance and have become the research focus of the field of aerospace, many approaches and tools have been recently developed and applied to different satellite missions, the noteworthy examples of these works are presented through an integrated framework with data mining methods. This study presents the state-of-art of the intelligent health monitoring operations and applications.

The remainder of this chapter is organized as follows; Sect. 2 presents an overview about satellite architecture, telemetry subsystem and telemetry data and its characteristics. Section 3 discuss the intelligent health monitoring system, shows main process and applications based on data mining and machine learning techniques. In addition, the conventional approached for health monitoring such as limit checking, expert system and model based diagnosis are presented. Finally, Sect. 4 presents the conclusion.

## 2 Satellite Telemetry Data

A satellite consists of a payload, which is the mission-specific equipment, and a collection of subsystems is called bus. Satellite bus is a group of components that support a common function that required for each satellites regardless of its mission. There is a difference between the payload and the rest of the satellite bus, because the payload is typically unique for a given mission, whereas the bus may be able to support different missions. A bus typically consists of Power Subsystem (PS), Communication

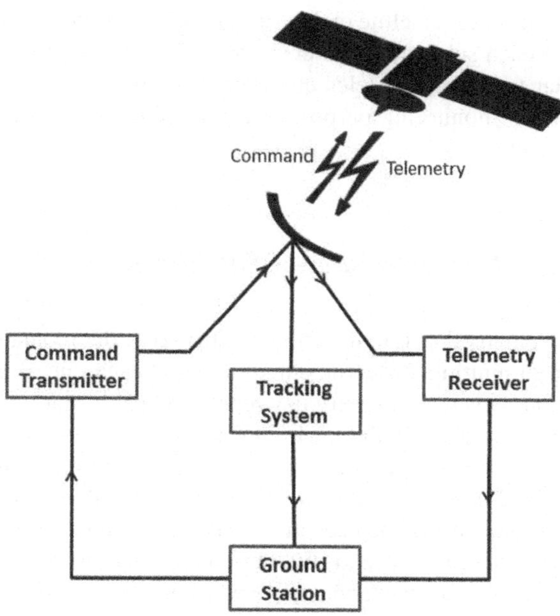


**Fig. 1** Main functions of Telemetry, Tracking and Command (TT&C) subsystem

Subsystem (CS), Attitude Determination and Control Subsystem (ADCS), Thermal Control Subsystem, Structure Subsystem, Command and Data Handling Subsystem (CDHS), Telemetry, Tracking and Command Subsystem (TT&C) [1–3].

TT&C is the main subsystem of the satellite bus that is considered as the main communication channel between the satellite and ground station for control and monitoring operations through collecting and sending telemetry data from all subsystems of the satellite to the ground station, and receiving transmitted commands from ground station. TT&C performs three main tasks; health monitoring operations is the first task, which is being achieved in the ground station through analyzing the received telemetry data. Second task, tracking the satellite locations, third, the proper control for the satellite through receiving the transmitted commands from the ground station [4, 5] (Fig. 1).

Telemetry is the collection of measurements and onboard instrument readings required to deduce the health and status of all of the satellite subsystems in the spacecraft bus and the payload. The TT&C subsystem must collect, process, and transmit this data from the satellite to the ground. It is a very important task in the operation of artificial satellites to monitor the health state of the system and detect any abnormal behavior.

During the artificial satellite operational lifetime the ground station is receiving the telemetry data, is non-stationary time series dataset contains thousands of sensor measurements from various subsystems, which contains the wealth information related to the health and status of the entire satellite and all its subsystems which reflect the operational status and payload of satellites. The health and status measurements of the satellite include the status of resources, the health and mode of operation for each subsystem and environmental data like values of sun and radiation or like star trackers. The telemetry data is analyzed in the ground control station for the health monitoring purposes such as failure diagnostic or prognostic, and anomaly detection [6, 7].

### *2.1 Characteristics of Satellite Telemetry Data*

Telemetry data is non-stationary time series data contains thousands of sensor outputs from multiple different sub-system and each one of these subsystems brings up to thousands records every day represent health, status and mode of each one, besides the thousands of the environmental changes and attitude of the satellite measurements. Telemetry data is a heterogeneity and multi-modality data, since it is being composed of hundreds to thousands of variables and attributes, are collected from various sensors and each one has different output form. Also each one of the satellite subsystem has several different operational modes and changes from one mode to another over time, and each one has different structure and parameters [6, 8].

## 3 Health Monitoring System

Spacecraft health monitoring is essential to ensure that a satellite is operating properly and has no anomalies that could threaten its mission. Any satellite needs monitoring, analyzing and controlling with respect to the requirement, detection, diagnostic and predication are the three main processes for monitoring the functions and behavior of the satellite and to ensure that it is operating properly and keep its performance.

### *3.1 Health Monitoring Based on Conventional Techniques*

Conventionally, detection and diagnosis have been proposed based on prior expert knowledge and deductive reasoning process such as expert systems and model-based diagnosis. In this section, a review of the three conventional approaches limit checking, expert system and model based diagnosis have been introduced.

#### 3.1.1 Limit Checking

Limit checking is the easiest and fastest method has been widely used in anomaly detection of spacecraft systems, it is based on monitoring upper and lower limits that have been set by the experts for various sensor values, such voltage, current, capacity, temperature, velocity, and so on. The generic check system is the first limit checking system for detecting anomalies of spacecraft was developed and applied by NASA in 1980 [9, 10]. However, the limit checking method has some limitations, such as the difficulty of identifying and modifying threshold values for the different conditions and modes, also if the limit values are not appropriate, they will either fail to detect any abnormalities at all or produce a large number of false alarms, making operators insensitive to the real anomalies [7]. To overcome these limitations, a number of adaptive limit checking have been developed, as the introduced work in [11] that has been proposed using regression tree learning algorithm to adapt limits of telemetry measurement in an automatic way for adoptively predicting the upper and lower limits of each sensor measurement, the experimental results demonstrated that its effectiveness when applied on archived telemetry data of artificial satellite. Based on intelligence learning algorithms there are more works have been developed for anomaly detection, as Relevance Vector Machine (RVM) which was originally introduced by Tipping [12], it has been applied to detect anomalies of real time telemetry data by obtaining predictive model [7], the model has been experimented on a real satellite telemetry data provided from Japan Aerospace Exploration Agency (JAXA), that has been demonstrated the model effectiveness for detecting different types of anomalies or failures quickly. Kernel Principal Component Analysis (Kernel PCA) is the another example of adaptive limit checking works has been applied for detecting anomalies in real satellite telemetry data from JAXA satellite [13] and proofed its effectiveness to predict the upper and lower limits and then detect anomalies in effective way. Though the effectiveness of adaptive limit checking, but there still exist a number of anomalies types occur without violating upper or lower limits of sensor values, so that cannot be detected just by monitoring the limits on the variables.

### 3.1.2 Expert Systems

In the early 1980s, many works of anomaly detection for space systems based on experience based method were studied and developed. The first work of expert system is designed as the classification problem conducted by the knowledge database that was previously prepared by domain experts [14, 15]. Another example of expert system has been developed by Japanese Institute of Space and Astronautically Science (ISAS) at the end of 90s, called "ISAACS-DOC (Intelligent Satellite Control Software-DOCtor)" to perform the safe operation of the spacecraft control on the ground with small number of operators, it has been used to geomagnetic observation satellite "GEOTAIL" for almost 10 years. Then, followed by the second version of ISAACS-DOC, which is the fully automated and requires less operator ground support system, which has been applied successfully of detecting anomalies for a series of spacecraft missions such as GEOTAIL, NOZOMI and HAYABUSA [15–17].

Though, the expert system is more effective than the limit checking in anomaly detection, but the ability of detecting anomalies is limited to the knowledge is described in advance which makes it is unable to deal with "unknown" anomalies.

### 3.1.3 Model-Based Diagnosis

Model-based diagnosis is another technique to detect anomalies that cannot be detected by the limit checking, it is based on comparing the observed behavior of the target systems with the simulation results obtained from some computational models [18]. The Livingstone is based on model-based diagnosis system that applied on NASA's DeepSpaceOne (DS-1) mission, the Livingstone system is used to monitor commands executions, failure detection and recovery, from using the model of the spacecraft's components and the command stream to predict the values of the sensors that should result from the commands assuming no components are failed. If there is a discrepancy between the predicted and observed sensors, then a failure is detected [19, 20].

Some of recent work based on qualitative method, such as; in [21] the authors had proposed a mixed architecture to face autonomic failure diagnosis within the space domain application, by mixing two different techniques of detection and identification, driven by a qualitative knowledge base of the system to be monitored. The Detection mechanism is based on the inductive reasoning approach, supported by the fuzzy logic theory to deal with uncertainties intrinsic in data coming from sensor readings. The proposed approach was tested on real space scenario: the GOCE spacecraft—An European Space Agency project carried on by Alenia Spazio in Turin, the results demonstrated that the failure detection & identification of real failures related to the Electric Power Subsystem are successfully managed.

Though the model-based diagnosis methods are effective and flexible, it has difficult to acquire accurate and complete models of space systems. The Conventional approaches are generally heavily dependent on a priori knowledge on the system behavior and on the knowledge of experts. As that is discussed for the model-based method where requires a perfect dynamics model of the system behavior, and the expert system demands a set of production rules. In practice, preparing such complete and accurate a priori knowledge of the systems is very difficult because of the difference between the simulated and actual behavior of the system on the orbit. As well as, due the inherent properties of telemetry data have been discussed above, the conventional methods are not sufficient enough for analyzing and extracting embedded information from telemetry data for monitoring the health state of space systems.

## *3.2 Intelligent Health Monitoring Based on Data Mining Techniques*

All conventional approaches are deductive, and cannot use the history spacecraft telemetry data effectively. Meanwhile the alternative approaches to the detection and diagnostic problems based on data mining techniques are recently developed and get more and more attention. The basic idea of these approaches is to introduce the

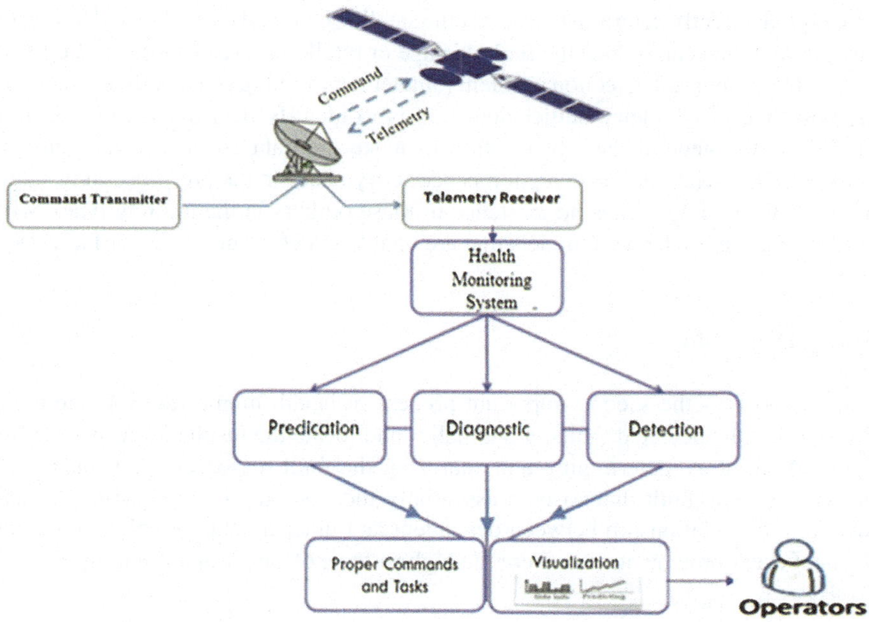


**Fig. 2** Main process and results of intelligent health monitoring system

system behavior models necessary for detection and diagnosis based on the historical data automatically or semi-automatically, rather than from the expert knowledge.

As shown in Fig. 2, detection, diagnostic and predication are the three main processes for monitoring the functions and behavior of the satellite and to ensure that it is operating properly and keep its performance. Results of the health monitoring system will be translated to the proper commands and tasks are sent to the satellite for reconfiguration or reschedule new task, also the results of the monitoring are visualized to provide the operators in the ground control station with useful information by summarizing the large amount of data, to assist for understanding the health status and detecting anomalies or failures of the in-orbit satellite, to take action as soon as possible. All previous processes will lead to the proper control of satellite based on the status of resources and mission operations.

### 3.2.1 Detection

Many detection problems have been addressed for telemetry data. Anomaly detection is one of the most important problem for telemetry mining, where the most of anomalies in data refer to significant and critical information. Therefore, early detection of anomalies is a critical task of health monitoring for satellites to avoid serious faults such as loss of control. Fault detection is another important detection process that directly determines whether the satellite can safety, reliable and long-life operation, which can reduce the fault damage or totally failure of in-orbit satellites.

Another important detection problem for telemetry mining is the outlier detection, the objective in telemetry outlier detection is to identify data objects that do not fit well in the general data distribution to make the data consistent and remove outliers. The trivial outliers detection is necessary phase in the pre-processing stage of telemetry mining, since the existence of these outliers in the training phase will produce inaccurate learned models that are unable to detect anomalies or faults [8].

### 3.2.2 Diagnostic

The diagnosis is the second important process of health monitoring for satellites, to diagnosis the detected faults or anomalies and locate the resulted effects of these faults. As well as, determining and analyzing the fault reason and the impact on the system. The fault diagnosis is essentially the first step of predication, which discovers the relationship between dependent and independent variables to predict the next event or behavior to ensure the highest level of safety and reliability.

### 3.2.3 Prediction

Prediction using telemetry data mainly aims at predicting the satellite subsystems status and performance in the future. Predication can be framed as a supervised learning problem, which discovers the relationship between independent variables and the relationship between dependent and independent variables to predict the next event or behavior. The objective of most of the studies on telemetry based prediction is to predict the faults, which is one of the key technologies for health monitoring of satellites, but there are some studies like [22, 23], which aim at prediction and estimation of the remaining usage life based on changing trends of the performance of space systems, which is the important basis of the realization of the system prognostics and health management.

### 3.2.4 Telemetry Visualization

The purpose of previous mining techniques was to develop telemetry monitoring systems that detect anomalies and predict the failures. Another promising application of telemetry data mining technology is the visualization that used for summarizing the large amount of data, to assist for understanding the health status and detecting anomalies or failures of the in-orbit satellite. Data visualization is defined as a modern branch of descriptive statistics that used to help the operators in analyzing the stream of received telemetry data. The main goal of visualization is to present information in efficiency ways to users via graphical tables, plots and charts [24, 25]. Telemetry visualization is being proposed for supporting a satellite operator's acquiring knowledge about relation of sequences, by detecting the changes of behavior of sequences, which helps operators to find the anomaly symptoms, and handle properly as soon as possible, many approaches of telemetry visualization have been recently developed and applied efficiently such as have been discussed in [26–28].

## 3.3 Intelligent Health Monitoring Applications

In the research community of data mining, a number of researchers are increasingly interested in applying data mining technology to the health monitoring problems for space missions, the following figure shows the major application problems of satellite health monitoring based on data mining and machine learning techniques.

As shown in Fig. 3, the main health monitoring processes can be classified in three classes; detection, predication and diagnosis, which are achieved by early detection for any abnormal behavior that can be lead to failure or loss of control such as anomalous symptoms [6, 7, 29–40], fault [41–49] or outlier detection [8, 50]. As well as, the diagnostic and prediction of faults through using real-time and historical state information of subsystems can lead to appropriate action to be scheduled proactively to avoid catastrophic failures [6, 7, 51–61]. In addition to, analyzing and assessing risk and reliability in systems for the purpose of improving safety and performance based on some features, such as the spare components, the dependent failures, common cause failure and the failure recoveries. The failure recovery is important issue to monitor and control the health of the satellite since the satellite fulfils their mission in a very challenging environment which is difficulty to eliminate the possibility of the sensor failure and lose the measurements [62–67].

The harsh environment also raises the importance of the simulation application, which is proposed to simulate accurately subsystem performance under the lifecycle conditions like environmental hazards and dependent actions, which are important features that must be considered during satellite design, construction, and safety assessment. One of the most important problem of the diagnosis is analyzing and extracting hidden relationship between apparently unrelated telemetry parameters using association rule mining as that used in [41, 68]. As well as, the Remaining Useful Life (RUL) Estimation is considered as an important application of health monitoring based on telemetry mining, for predicting the remaining useful life of a sensor or subsystem given the current status, historical and coming loads and environmental conditions [69–73].

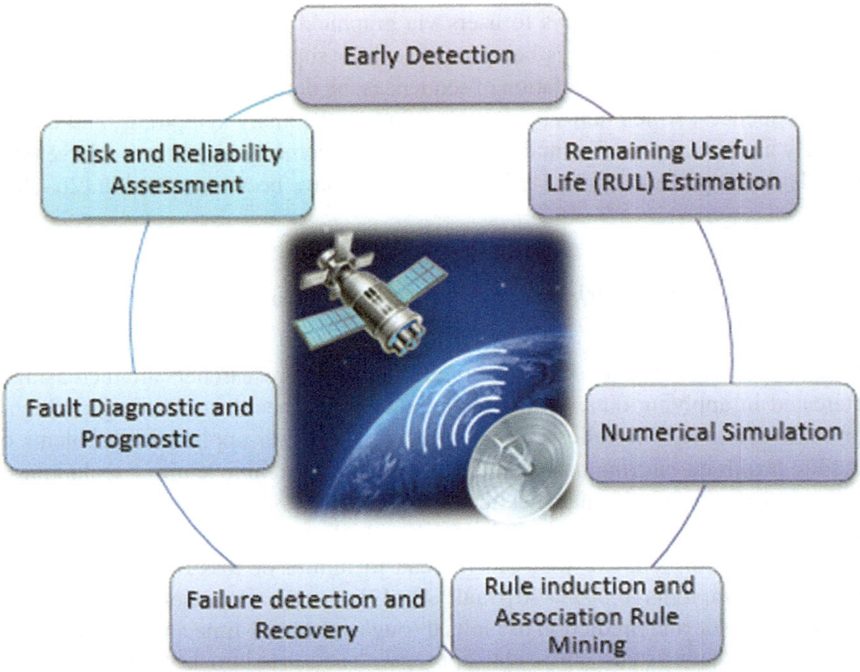


**Fig. 3** Main applications of intelligent health monitoring system

## 4 Conclusion

Satellite telemetry data is non-stationary time series data contains thousands of sensor outputs from multiple different sub-systems contains the wealth information related to the health and status of the satellite and actual operating state of each subsystem, the stream of these health information is analyzed in the ground control station for routine operational and failure diagnostic purposes. This chapter reviewed the literature to identify the health monitoring problems that have been solved by telemetry data mining and machine learning techniques. It further surveyed group of application problems with the corresponding mining methods. The major contribution of this chapter is in providing an integrated view about space systems and ground control operations, through the telemetry data mining application problems with the methods used to address them. The work presented in this study can be used as a guideline for designing intelligent health monitoring solution with a suitable mining method.

**Acknowledgements** This work is supported by Egypt Knowledge and Technology Alliance (E-KTA) for Space Science, which is supported by The Academy of Scientific Research & Technology (ASRT), and coordinated by National Authority for Remote Sensing and Space Sciences (NARSS) (TEDDSAT Project grant).

## References


1. L. Zhou, A. Junshe, Design of a payload data handling system for satellites, in *Third International Conference on Instrumentation, Measurement, Computer, Communication and Control (IMCCC)*, IEEE, Shenyang, China (2013)
2. A. Nicolai, S. Roemer, S. Eckert, The TET satellite bus—future mission capabilities, in *Aerospace Conference*, IEEE, Big Sky, MT, USA (2014)
3. S. Roemer, S. Eckert, The TET satellite bus-a high reliability bus for LEO applications, in *28th International Symposium on Space Technology and Science*, Okinawa (2011)
4. B. Anyaegbunam, Design elements of satellite telemetry, tracking and control subsystems for the proposed nigerian made satellite. Int. J. Eng. Sci. Inven. **3**(1), 5–13 (2014)
5. P.K. Udaniya, G. Sharma, L. Tharani, Application of MIMO system for telemetry, tracking command and monitoring subsystem to control the satellite, in *International Conference on Computing, Communication and Automation (ICCCA2016)*, IEEE, Greater Noida, India (2016)
6. L. Quan, Z. XingShe, L. Peng, L. Shaomin, Anomaly detection and fault diagnosis technology of spacecraft based on telemetry-mining, in *2010 3rd International Symposium on Systems and Control in Aeronautics and Astronautics (ISSCAA)*, IEEE, Harbin, China (2010)
7. T. Yairi, Y. Kawahara, R. Fujimaki, Y. Sato, K. Machida, Telemetry-mining: a machine learning approach to anomaly detection and fault diagnosis for space systems, in *2nd IEEE International Conference on Space Mission Challenges for Information Technology*, IEEE, CA, USA (2006)
8. T. Yairi, N. Takeishi, T. Oda, Y. Nakajima, N. Nishimura, N. Takata, A data-driven health monitoring method for satellite housekeeping data based on probabilistic clustering and dimensionality reduction. IEEE Trans. Aerosp. Electron. Syst. **53**(3), 1384–1401 (2017)
9. G. Wang, L. Qiang, S. Jinglin, M. Xiaofeng, Telemetry data processing flow model: a case study. Aircr. Eng. Aerosp. Technol. Int. J. **87**(1), 52–58 (2015)
10. R. Fujimaki, T. Yairi, K. Machida, Adaptive limit-checking for spacecraft using relevance vector autoregressive model, in *8th International Symposium on Artificial Intelligence, Robotics and Automation in Space—iSAIRAS, ESA SP-603*, Munich, Germany (2005)
11. T. Yairi, M. Nakatsugawa, K. Hori, S. Nakasuka, K. Machida, N. Ishihama, Adaptive limit checking for spacecraft telemetry data using regression tree learning, in *2004 IEEE International Conference on Systems, Man and Cybernetics*, IEEE, The Hague, Netherlands (2004)
12. M. Tipping, Sparse bayesian learning and the relevance vector machine. J. Mach. Learn. Res. **1**, 211–244 (2001)
13. I. Minoru, K. Yoshinobu, G. Kohei, Y. Takehisa, M. Kazuo, Adaptive limit checking for spacecraft telemetry data using kernel principal component analysis. Trans. Jpn. Soc. Aeronaut. Space Sci. Space Technol. Jpn. **7**(26), 11–16 (2009)
14. C. Chang, W. Nallo, R. Rastogi, D. Beugless, F. Mickey, A. Shoop, Satellite diagnostic system: an expert system for intelsat satellite operations, in *IVth European Aerospace Conference (EAC)* (1992), pp. 321–327
15. N. Nishigori, M. Hashimoto, A. Choki, M. Mizutani, Fully automatic and operator-less anomaly detecting ground support system for mars probe 'NOZOMI', in *6th International Symposium on Artificial Intelligence and Robotics and Automation in Space (I-SAIRAS)* (2001)
16. M. Hashimoto, N. Nishigori, M. Mizutani, Running status of monitoring and diagnostic expert system for mars observer "NOZOMI", in *The 22nd International Symposium on Space Technology and Science, ISTS* (2000)
17. M. Hashimoto, N. Nishigori, M. Mizutani, Operating status of monitoring and diagnostic expert system for geomagnetic satellite GEOTAIL, in *The 2nd International Symposium on 'Reducing the Cost of Spacecraft Ground Systems and Operation'* (1997), pp. 1–8
18. B.C. Williams, P.P. Nayak, A model-based approach to reactive self-configuring systems, in *The Thirteenth National Conference on Artificial Intelligence* (1996), pp. 971–978
19. J. Kurien, M. Dolores, Costs and benefits of model-based diagnosis, in *Aerospace Conference*, IEEE, MT, USA (2008)
20. S.C. Hayden, A.J. Sweet, S.E. Christa, Livingstone model-based diagnosis of earth observing one, in *AIAA 1st Intelligent Systems Technical Conference*, CA, United States (2004)

21. A.E. Finzi, M.R. Lavagna, G. Sangiovanni, Fuzzy inductive reasoning and possibilistic logic for space systems failure smart detection and identification, in *The 7th International Symposium on Artificial Intelligence, Robotics and Automation in Space: i-SAIRAS 2003*, NARA, Japan (2003)
22. H. Fang, Y. Xing, K. Luo, H. Liming, Study of the long-term performance prediction methods using the spacecraft telemetry data, in *Prognostics & System Health Management Conference (PHM-2012)*, IEEE, Beijing (2012)
23. C. Sary, C. Peterson, I. Rowe, T. Ames, Trend analysis for spacecraft systems using multimodal reasoning, in *AAAI Spring Symposium*, Technical Report (2008), pp. 152–158
24. J. Wijk, E. Selow, Cluster and calendar based visualization of time series data, in *IEEE Symposium on Information Visualization (InfoVIs. I 99)*, IEEE, San Francisco, California (1999), pp. 4–9
25. D.A. Keim, Information visualization and visual data mining. IEEE Trans. Vis. Comput. Graph. **8**(1), 1–8 (2002)
26. X. Dong, P. Dechang, An effective method for mining quantitative association rules with clustering partition in satellite telemetry data, in *2014 Second International Conference on Advanced Cloud and Big Data*, IEEE, Huangshan, China (2014), pp. 26–35
27. J. Lin, E. Keogh, S. Lonardi, J.P. Lankford, D.M. Nystrom, VizTree: a tool for visually mining and monitoring massive time series databases, in *The 30th VLDB Conference*, Toronto, Canada (2004)
28. Y. Gao, Y. Tianshe, X. Minqiang, N. Xing, An unsupervised anomaly detection approach for spacecraft based on normal behavior clustering, in *Fifth International Conference on Intelligent Computation Technology and Automation*, Zhangjiajie, China (2012)
29. D.D. Coste, M.B. Levine, Automated event detection in space instruments: a case study using IPEX-2 data and support vector machines, in *SPIE Conference Astronomical Telescopes and Instrumentation* (2000)
30. R. Fujimaki, T. Yairi, K. Machida, An anomaly detection method for spacecraft using relevance vector learning, in *Advances in Knowledge Discovery and Data Mining, 9th Pacific-Asia Conference*, Springer, Hanoi, Vietnam (2005)
31. D.L. Iverson, System health monitoring for space mission operations, in *Aerospace Conference*, IEEE, MT, USA (2008)
32. D. Azevedo, A. Ambrósio, M. Vieira, Applying data mining for detecting anomalies in satellites, in *Ninth European Dependable Computing Conference (EDCC)*, IEEE Computer Society, Sibiu, Romania (2012), pp. 212–217
33. X. Bing, L. Zhan, An anomaly detection method for spacecraft using ICA technology, in *ICACSEI* (2013)
34. B. Nassar, W. Hussein, M. Mokhtar, Space telemetry anomaly detection based on statistical PCA algorithm. Int. J. Electron. Commun. Eng. **9**(6) (2015)
35. L. Jin, M. Huang, Y. Jingjing, The anomaly mixed spectrum signals detection based on ICA and KNN. DEStech Trans. Eng. Technol. Res. (2016)
36. L. Datong, W. Shaojun, J. Chen, J. Zhou, Y. Peng, Anomaly detection with improved similarity measure for satellite telemetry data, in *Proceedings of the 22nd ISSAT International Conference on Reliability and Quality in Design*, International Society of Science and Applied Technologies, California, USA (2016)
37. B. Nassar, W. Hussein, Statistical learning approach for spacecraft systems health monitoring, in *Proceeding 2016 IEEE Aerospace Conference*, IEEE, MT, USA (2016)
38. B. Gautam, H. Khorasgani, G. Stanje, A. Dubey, D. Somnath, S. Ghosha, An approach to mode and anomaly detection with spacecraft telemetry data. Int. J. Progn. Health Manag. 1–18 (2016)
39. L. Datong, P. Jingyue, G. Song, X. Wei, Y. Peng, P. Xiyuan, Fragment anomaly detection with prediction and statistical analysis for satellite telemetry. IEEE Access **5**, 19269–19281 (2017)
40. M.M. Fernández, Y. Yisong, W. Romann, Telemetry anomaly detection system using machine learning to streamline mission operations, in *Proceeding 6th IEEE International Conference on Space Mission Challenges for Information Technology*, IEEE Computer Society, Palo Alto, California (2017), pp. 70–75

41. T. Yairi, K. Yoshikiyo, H. Koichi, Fault detection by mining association rules from house-keeping data, in *International Symposium on Artificial Intelligence, Robotics and Automation in Space* (2001)
42. L.B. Jack, A.K. Nandi, Fault detection using support vector machines and artificial neural networks, augmented by genetic algorithms. Mech. Syst. Signal Process. J. **16**(2–3), 373–390 (2002)
43. Y. Zhang, Fault detection and diagnosis of nonlinear processes using improved kernel independent component analysis (KICA) and support vector machine (SVM). Ind. Eng. Chem. Res. J. **47**(18), 6961–6971 (2008)
44. T. Bhekisipho, Predicting software faults in large space systems using machine learning techniques. Def. Sci. J. **61**(4), 306–316 (2011)
45. Y. Gao, T. Yang, N. Xing, Fault detection and diagnosis for spacecraft using principal component analysis and support vector machines, in *2012 7th IEEE Conference on Industrial Electronics and Applications (ICIEA)*, IEEE, Singapore, Singapore (2012)
46. T. Yang, B. Chen, H. Zhang, X. Wang, Y. Gao, N. Xing, State trend prediction of spacecraft based on BP neural network, in *2013 2nd International Conference on Measurement, Information and Control*, Harbin, CHINA (2013)
47. R. Wang, X. Gong, X. Minqiang, L. Yuqing, Fault detection of flywheel system based on clustering and principal component analysis. Chin. J. Aeronaut. **28**(6), 1676–1688 (2015)
48. P.K. Ray, B.K. Panigrahi, P.K. Rout, A. Mohanty, H. Dubey, Detection of faults in power system using wavelet transform and independent component analysis, in *Proceeding First International Conference on Advancement of Computer Communication & Electrical Technology*, Murshidabad, India (2016)
49. J. Carvajal, G. Jian, G. Eberhard, Agent-based algorithm for fault detection and recovery of gyroscope's drift in small satellite missions. Acta Astronaut. **139**, 181–188 (2017)
50. F. Bouleau, S. Christoph, Towards the identification of outliers in satellite telemetry data by using fourier coefficients, in *Proceedings of 6th International Conference on Agents an Artificial Intelligence*, Angers, France (2014), pp. 211–224
51. Z. Al-Dein, K. Khorasani, Neural network-based actuator fault diagnosis for attitude control subsystem of an unmanned space vehicle, in *Proceeding International Joint Conference on Neural Networks*, IEEE, Piscataway (2006), pp. 3686–3693
52. F. Song, V.Z. Cheng, Exploring event correlation for failure prediction in coalitions of clusters, in *SC '07 Proceedings of the 2007 ACM/IEEE Conference on Supercomputing*, New York, USA (2007)
53. J. Schumann, O. Mengshoel, T. Mbaya, Integrated software and sensor health management for small spacecraft, in *Proceeding the Fourth International Conference on Space Mission Challenges for Information Technology (SMC-IT)*, IEEE, Palo Alto, California (2011), pp. 77–84
54. Y. Gao, T. Yang, J. Feng, X. Minqiang, A neural network approach for satellite telemetry data prediction, in *ICECC '12 Proceedings of the 2012 International Conference on Electronics, Communications and Control*, Washington, USA (2012)
55. Y. Tianshe, B. Chen, Y. Gao, F. Junhua, H. Zhang, X. Wang, Data mining-based fault detection and prediction methods for in-orbit satellite, in *Proceeding International Conference on Measurement, Information and Control (ICMIC)*, IEEE, Harbin, China (2013)
56. S. Xie, X. Peng, X. Zhong, C. Liu, Fault diagnosis of the satellite power system based on the Bayesian network, in *Proceedings of the 8th International Conference on Computer Science and Education*, IEEE, Piscataway (2013), pp. 1004–1008
57. G. Xiang, T. Zhang, J.L. Hong, G. Jian, Spacecraft fault diagnosis based on telemetry data mining and fault tree analysis and design of expert system. Adv. Mater. Res. **760**, 1062–1066 (2013)
58. I. Gueddi, O. Nasri, K. Benothman, D. Philippe, VPCA-based fault diagnosis of spacecraft reaction wheels, in *Proceeding International Conference on Information, Communication and Automation Technologies (ICAT)*, IEEE, Sarajevo, Bosnia, Herzegovina (2015)
59. O. Nasri, I. Gueddi, D. Philippe Dague, B. Kamal, Spacecraft actuator diagnosis with principal component analysis: application to the Rendez-Vous phase of the mars sample return mission. J. Control Sci. Eng. **2015**, 1–11 (2015)

60. Y. Mounir Yassin, A. El-Mahallawy, A. El-Sharkawi, Real time prediction and correction of ADCS problems in LEO satellites using fuzzy logic. Egypt. J. Remote Sens. Space Sci. **20**, 11–19 (2017)
61. S. Skobtsov, N. Novoselova, V. Arhipov, S. Potryasaev, Intelligent telemetry data analysis of small satellites, in *The 6th Computer Science On-line Conference 2017 (CSOC2017)*, vol. 2 (2017), pp. 351–361
62. A. Guiotto, A. Martelli, C. Paccagnini, M. Lavagna, SMART-FDIR: use of artificial intelligence in the implementation of a satellite FDIR, in *Data Systems in Aerospace (DASIA)* (2003)
63. R. Gessner, B. Kosters, A. Hefler, R. Eilenberger, J. Hartmann, M. Schmidt, Hierarchical FDIR concepts in S/C systems, in *Proceedings of the 8th International Conference on Space Operations (SpaceOps)*, AIAA (2004), pp. 233–249
64. L. Portinale, R.D. Codetta, S. Di Nolfo, A. Guiotto, ARPHA: a software prototype for fault detection, identification and recovery in autonomous spacecrafts. Acta Futur. **5**, 99–110 (2012)
65. A. Zolghadri, Advanced model-based FDIR techniques for aerospace systems: today challenges and opportunities. Prog. Aerosp. Sci. **53**, 18–29 (2012)
66. A. Wander, R. Forstner, Innovative fault detection, isolation and recovery on-board spacecraft: study and implementation using cognitive automation, in *Proceeding Conference on Control and Fault-Tolerant Systems (SysTol)*, IEEE, Piscataway (2013), pp. 336–341
67. S. Abdelghafar, A. Darwish, A.E. Hassanien, Cube satellite failure detection and recovery using optimized support vector machine, in *Proceedings of The International Conference on Advanced Intelligent Systems and Informatics*, Cairo, Egypt (2018), pp. 664–674
68. S.A. Kannan, T. Devi, Mining satellite telemetry data: comparison of rule-induction and association mining techniques, in *IEEE International Conference on Advances in Computer Applications (ICACA)*, IEEE, Tamil Nadu, India (2016), pp. 259–264
69. S. Bhaskar, K. Goebel, S. Poll, J. Christophersen, Prognostics methods for battery health monitoring using a Bayesian framework. IEEE Trans. Instrum. Meas. **58**(2), 291–296 (2009)
70. S. Bhaskar, K. Goebel, J. Christophersen, Comparison of prognostic algorithms for estimating remaining useful life of batteries. Trans. Inst. Meas. Control **31**(4), 293–308 (2009)
71. J.A.M. Penna, C.L. Nascimento, L.R. Rodrigues, Health monitoring and remaining useful life estimation of lithium-ion aeronautical batteries, in *Processing 2012 IEEE Aerospace Conference* (2012), Big Sky, MT, USA, pp. 1–12
72. L. Datong, P. Jingyue, Z. Jianbao, P. Yu, Data-driven prognostics for lithium-ion battery based on Gaussian process regression, in *Proceeding 2012 IEEE Conference on Prognostics and System Health Management (PHM)*, IEEE (2012), pp. 1–5
73. Y. Jinsong, M. Baohua, T. Diyin, L. Hao, W. Jiuqing, Remaining useful life prediction for lithium-ion batteries using a quantum particle swarm optimization-based particle filter. Qual. Eng. J. **29**, 536–546 (2017) (Special Issue on Reliability Engineering)

# Design, Implementation, and Validation of Satellite Simulator and Data Packets Analysis

**Kadry Ali Ezzat, Lamia Nabil Mahdy, Aboul Ella Hassanien and Ashraf Darwish**

**Abstract** The objective of the communication subsystem is to communicate with ground stations to download information and transfer directions. The carrier-to-noise ratio of both the telemetry downlink and the order uplink is determined as a figure of legitimacy for the station conveying capacity of the connection. The proposed subsystem additionally enables the client to choose which ground stations are dynamic through a ground station menu. Alternate parameters in this menu are the ground station: name, scope, longitude, and elevation. Presently, there are 6 stations characterized. The client can include or erase from this rundown through this menu. The section isolated into three stages, stage 1 is to process the azimuth edge, rise point and separation among satellite and the ground station while stage 2 is to register the uplink and down connection parameters. Stage 3 is an isolated work and it manages crating query tables for information bundles.

## 1 Introduction and Basics

The interchanges subsystem interfaces the satellite with the ground or other satellite. Data sent to the satellite (i.e. uplink or forward connection), comprises of directions and required information to satellite (i.e. satellite control directions and new SW adaptation). Data got from the satellite (i.e. downlink or return connect) comprises of satellite status telemetry and payload information. The fundamental correspondence subsystem comprises of a recipient, a transmitter, and a wide-edge (hemispheric or

K. A. Ezzat (✉) · L. N. Mahdy
Biomedical Engineering Department, Higher Technological Institute, Cairo, Egypt
e-mail: Kadry_ezat@hotmail.com

A. E. Hassanien
Faculty of Computers and Information, Cairo University, Cairo, Egypt

A. Darwish
Faculty of Science, Helwan University, Helwan, Egypt

K. A. Ezzat · L. N. Mahdy · A. E. Hassanien · A. Darwish
Scientific Research Group in Egypt (SRGE), Cairo, Egypt



A. E. Hassanien et al. (eds.), *Machine Learning and Data Mining in Aerospace Technology*, Studies in Computational Intelligence 836,
https://doi.org/10.1007/978-3-030-20212-5_5

omni-directional) radio wire. Frameworks with high information rates may likewise utilize a directional receiving wire [1].

The essential capacity of the proposed SIM programming is to make a virtual condition to reproduce a rocket. The reenactment incorporates the shuttle's task and the association of different subsystems as a component of time and assets. The proposed SIM presents this virtual condition between the satellite and the ground. The Proposed satellite test system subsystem is approved by different recreations for independent locally available dispatch and early circle stage activities, oddity task, and science fine mode activity. It is additionally formally checked by effectively breezing through different tests, for example, the satellite test system subsystem test, mission control component framework combination test, interface test, site establishment test, and acknowledgment test [2].

The proposed test system characterizes a pecking order of square outlines wired together alongside parameters that portray operational and execution qualities that yields an all around recorded utilitarian shuttle demonstrate. Graphical UI (GUI) window characterizes a capacity or a chain of command of lower level squares. Squares at the least dimension conjure MATLAB® or SIMULINK® code. The GUI shows a discourse box to the client that enables changes to be made to a square's parameters previously reproduction begins. Lines associating the squares transmit qualities, for example, those used to speak to orbital data and shuttle assets.

# 2 Satellite Simulator Design Phase

## *2.1 The Output of Communication Subsystems*

### 2.1.1 Azimuth and Elevation Angle for Satellite Tracking

It is necessary for the earth station to know where the satellite is in the circle. At that point the earth station build needs to ascertain a few edges to follow the satellite effectively. These points are called radio wire look edge. The look plots for the ground station radio wire are the azimuth and height edges required at the reception apparatus with the goal that it focuses straightforwardly at the satellite. With the geostationary circle, the circumstance is a lot easier than some other circle [3]. As the radio wire shaft width is exceptionally restricted and following component is required to adjust for the development of the satellite about the ostensible geostationary position. Three snippets of data that are expected to decide the look plots for the geostationary circle are:

a. Earth station latitude
b. Earth station longitude
c. Satellite orbital position

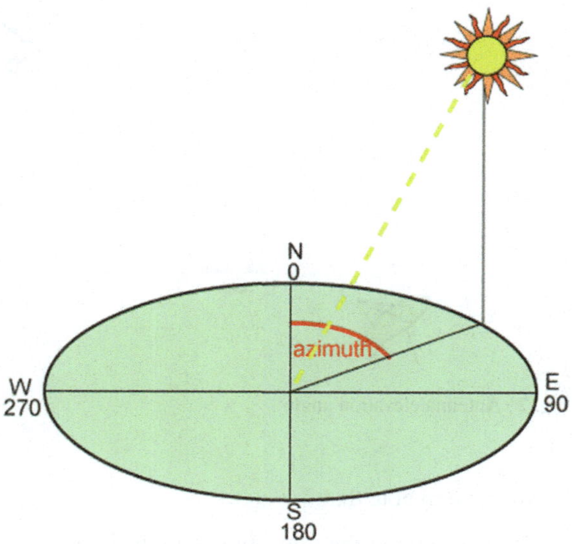


**Fig. 1** Antenna azimuth angle

Utilizing these data receiving wires look edge can be determined utilizing Napier's standard (illuminating round triangle). Azimuth point indicates the level edge estimated at the earth station receiving wire toward the north post. The rise is such edge means the vertical edge estimated at the earth station receiving wire end starting from the earliest stage satellite position as appeared in Fig. 1.

The equation for Azimuth (Az) determination is defined as follows.

$$\mathrm{Az} = 180^\circ + \tan^{-1}\left(\frac{\tan G}{\tan L}\right) \tag{1}$$

where:

*G* is *Difference between satellite orbital position and earth station antenna.*
*L* is *Latitude of your earth station antenna.*

In figure (1) Az means azimuth angle required to track the satellite horizontally. In Fig. 2 the elevation angle has been shown.

Equation for Elevation (El) determination

$$\mathrm{El} = \tan^{-1}\left(\frac{\cos G.\cos L - 0.1512}{\sqrt{1 - \cos^2 G.\cos^2 L}}\right) \tag{2}$$

where:

*G* is *Difference between satellite orbital position and earth station antenna.*
*L* is *Latitude of your earth station antenna.*
0.1512 *is constant.*

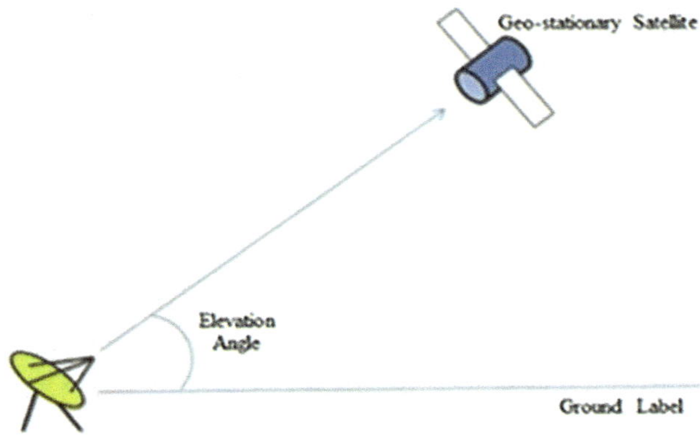


**Fig. 2** Antenna elevation angle

We have to note that:

1. If the satellite orbital location is in the east (E), then G = Antenna longitude – Satellite orbital position.
2. If the satellite orbital location is in West (W), then G = Satellite orbital position – Antenna longitude.

### 2.1.2 Distance Between Satellite and Earth Antenna

The place on the earth's surface where the dish antenna is located is denoted by P′. Assume that its position has spherical coordinates ($\lambda$, $\phi'$) where $\lambda$ always denotes longitude, measured positive east, and $\phi'$ indicates spherical (geocentric) latitude. Assume furthermore that the sub-satellite Point E intersection with the earth's surface of the geocentric radius vector to the satellite S is on the equator at a longitude $\lambda$s. The angle $\gamma$ between the radius vectors of points P′ and E can be obtained using the right spherical triangle P′ Q E. Applying Napier's rules it follows that

$$\cos\gamma = \cos\phi' \cos(\lambda_s - \lambda) \tag{3}$$

where:

$\phi'$ is indicates spherical (geocentric) latitude,
$\lambda_s$ is equator longitude,
$\lambda$ is longitude.

The distances OP′ and OS, respectively R and r, are related to the angle $\gamma$ by the equation

$$d = r[1 + (R/r)^2 - 2(R/r)\cos\gamma]^2 \tag{4}$$

where d—topocentric separate from the receiving wire to the satellite, or on the other hand, the scope of protest from eyewitness; r is utilized to signify the geocentric separation from the world's inside to the shuttle, which for a perfect geo-stationary satellite is consistent, r = 42,200 km; and R = a "mean esteem" for the range of the earth; the sweep of a circle that has indistinguishable volume from the earth ellipsoid; subsequently, R = 6371 km [1, 2].

### 2.1.3 Uplink and Downlink

The correspondence going from a satellite to ground is called downlink, and when it is going from ground to a satellite it is called uplink. At the point when an uplink is being gotten by the rocket in the meantime a downlink is being gotten by Earth, the correspondence is called two-way. In the event that there is just an uplink occurring, this correspondence is called transfer. In the event that there is just a downlink occurring, the correspondence is called one-way [4].

## 2.2 *Parameters of Simulator Inputs*

### 2.2.1 Antenna Gain

Isotropic power radiation is generally not compelling for satellite correspondences joins, in light of the fact that the power thickness levels will be low for generally applications. Some directivity (gain) is alluring for both transmit and get radio wires.

Look at first as a lossless (perfect) radio wire with a physical aperture area of A ($m^2$) [5, 6].

The gain of the perfect receiving device is defined as follows.

$$g_{ideal} = \frac{4\pi A}{\lambda^2} \tag{5}$$

where:

*A* is Physical aperture area,
λ is wave length.

So the ideal antennas are not practical, because some energy is reflected, some energy is absorbed by lossy components (feeds, struts, sub reflectors). To account for this, an effective aperture, $A_e$, is defined in terms of aperture efficiency.

$$A_e = \eta_e A \tag{6}$$

where:

*A* is Physical aperture area,

$\eta_e$ is aperture efficiency.

The physical antenna gain is denoted as G and computed by

$$G = 10\log\left[\eta_A \frac{4\pi A}{\lambda^2}\right] \tag{7}$$

where:

$A$ is Physical aperture area,
$\eta_e$ is aperture efficiency and
$\lambda$ is wave length.

### 2.2.2 Antenna Temperature

Antenna temperature ($T_A$) may be known if the total attenuation due to rain and gas absorption (A), the temperature of the rain medium ($T_m$) and the temperature of the cold sky ($T_C$) are also known. Then, the following expression may be applied:

$$T_A = T_m(1 - 10^{\frac{-A}{10}}) + T_c.10^{\frac{-A}{10}} \tag{8}$$

where:

$T_m$ is the temperature of the rain medium, $T_c$ is the temperature of the cold sky and $A$ is gas absorption.

Usually, for clouds, it is considered $T_m = 280$ K and for rain, $T_m = 260$ K.

### 2.2.3 System Temperature

Other components also provoke attenuation of the signal. In order to calculate $T_{comp}$ it is necessary to determine the effective noise temperature and the gain of each stage of the ground station receiver path, according to the Friis formula:

$$T_{comp} = T_1 + \frac{T_2}{G_1} + \frac{T_3}{G_1 G_2} + \frac{T_4}{G_1 G_2 G_3} \tag{9}$$

where:

$T_{1\to 4}$ is the temperature of each stage from $1 \to 4$,
$G_{1\to 3}$ is gain of each stage from $1 \to 3$.

### 2.2.4 Noise Temperature

The system noise temperature ($T_S$) is the sum of the antenna noise temperature ($T_A$) and the composite temperature of other components ($T_{comp}$) where,

$$T_S = T_A + T_{comp} \tag{10}$$

where:

$T_A$ is antenna noise temperature and
$T_{comp}$ is system temperature.

### 2.2.5 Noise Temperature of a Receiver

A perfect quiet accepting enhancer would intensify an information clamor not more than the info flag (i.e. with a similar gain). Because of inward clamor, a real accepting speaker will bring extra commotion control [7, 8].

The commotion caused by a recipient is normally expressed in terms of an equivalent amplifier noise temperature TR. It is characterized as the temperature of a commotion source (opposition) which, when associated with the contribution of a silent collector, gives indistinguishable clamor at the yield from the real recipient.

The collector is really made out of fell circuits and, all the more accurately of a couple intensifying stages or different systems (for example, the down-converter, and so forth.), every one having its own gain gi and its noise temperature $T_{Ri}$. It tends to be effectively shown that, under such conditions the collector commotion temperature is:

$$T_R = T_{R1} + (T_{R2}/g_1) + (T_{R3}/g_1 g_2) + \ldots \tag{11}$$

This recipe is critical in light of the fact that it demonstrates that the clamor commitments of the progressive stages are diminished by the aggregate gain of the previous stages. Consequently, the RF amplifier, called the low noise amplifier (LNA) must have a low $T_{R1}$ and a high $_{g1}$.

Common values of $T_R$ for the LNAs used in modern receivers are between 30 and 150 K, depending on the frequency band and on the LNA design.

Note that, in small earth stations (receive-only, small stations for business communications, called VSATs, etc.), the LNA is generally included with the down converter (D/C) in a single unit called low noise converter (LNC) or low noise block (LNB).

Note additionally that the commotion caused by the recipient is in some cases communicated by its clamor Figure F (or by FdB = 10 log F), the connection among F and $T_R$ being: TR = (F − 1) $T_0$, $T_0$ being, by tradition, equivalent to a typical surrounding temperature estimation of 290 K. Truth be told, since $T_R$ is normally substantially less than 290 K, $T_R$ is more viable to use than the clamor figure in satellite interchanges [9, 10].

### 2.2.6 Noise Temperature of an Antenna

The noise temperature of an antenna is the translating, in terms of noise temperature of the collection, by the antenna of the external noise [11].

### 2.2.7 Noise Temperature of a Receiving System

Figure 3 demonstrates a functional getting framework, with a reception apparatus with a commotion temperature $T_A$ and a collector with a clamor temperature TR.

A lessening area is embedded between the two sections. This area speaks to the misfortunes (for the most part ohmic misfortunes) in the reception apparatus and in the feeder (i.e., the RF interface, waveguide, coaxial or some other component) [10].

### 2.2.8 Equivalent Isotropic Radiated Power (EIRP)

EIRP considered as the information control in one end of the connection. EIRP is likewise presented unequivocally toward the start all things considered so it very well may be comprehended the wellspring of every segment and to permit the right appreciation of the considerable number of derivations displayed. The most extreme power motion thickness at a separation r is given by:

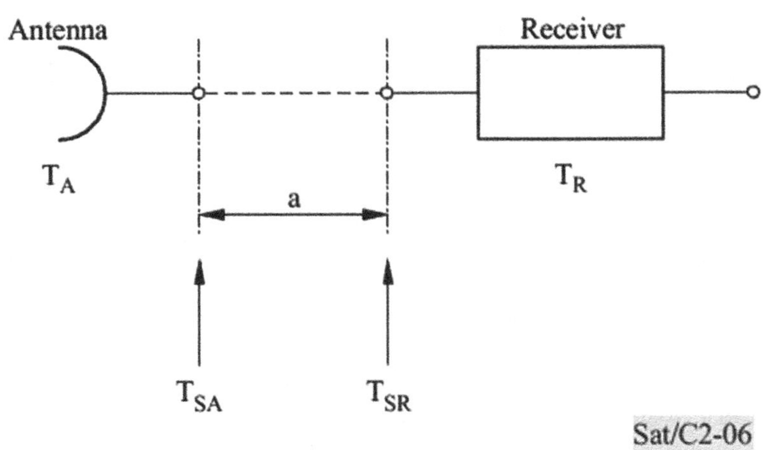


**Fig. 3** Noise temperature of a receiving system. **a** Is the attenuation, expressed as a power ratio ($a \geq 1$, i.e. in decibels, $a_{dB} = 10 \log a$). $\mathbf{T_a}$ is the physical temperature of the attenuating section (generally taken = 290 K). $\mathbf{T_R}$ is the physical temperature of the receiver. $\mathbf{T_{SR}}$ is referred at the receiver input, which means that, in subsequent calculations, the receiver can be assumed to be noiseless. $\mathbf{T_{SA}}$ is referred at the antenna output, which means that, in subsequent calculations, the attenuating section and the receiver can be assumed to be noiseless

$$\psi_M = \frac{G_T.P_S}{4\pi r^2} \tag{12}$$

where:

$\psi_M$ is maximum power flux density.
$G_T$ is transmission antenna gain.
$P_S$ is radiated power from the antenna.
$r$ is distance between the satellite and the receiving station.

Considering an isotropic radiator with an input power equal to G.$P_S$ the same flux density would be produced.

$$EIRP = G_T.P_S \tag{13}$$

where:

$G_T$ is transmission antenna gain.
$P_S$ is radiated power from the antenna.

Once EIRP is usually expressed in dBW, it is possible to write:

$$EIRP(dBW) = G_T(dBW) + P_S(dBW) \tag{14}$$

where:

$G_T$ is transmission antenna gain.
$P_S$ is radiated power from the antenna.

### 2.2.9 Free Space Losses (FSL)

Free space loss is the prevailing part in the loss of the strength of the signal. It doesn't related to attenuation of the signal, however to its spreading through space.

The first step in the calculations for free space loss (FSL) is to determine the losses in clear-sky conditions. These are the losses that remain constant with time. As said before, FSL derives from the spreading of the signal in space.

*FSL* is given by the following expression:

$$FSL = 10\log\left(\left(\frac{4\pi r f}{c}\right)^2\right) \tag{15}$$

where:

$r$ is distance,
$f$ is frequency and
$c$ is light speed.

### 2.2.10 The Figure of Merit *(G/T)*

$$\text{Figure of Merit} = \frac{G}{T_s} \tag{16}$$

where:

$G$ is Antenna gain of the receiver
$T_s$ is System noise temperature

is an Antenna gain-to-noise-temperature in the characterization of antenna performance, where G is the antenna gain in decibels at the receive frequency, and T is the equivalent noise temperature of the receiving system in kelvins. The receiving system noise temperature is the summation of the antenna noise temperature and the RF chain noise temperature from the antenna terminals to the receiver output [5].

### 2.2.11 Carrier to Noise Density Ratio ($C/N_0$)

($C/N_0$) is the ratio of the carrier power C to the noise power density $N_0$, expressed in dB-Hz. When considering only the receiver as a source of noise, it is called carrier-to-receiver-noise-density ratio:

$$\frac{C}{N_0} = \text{EIRP}\frac{G}{T}\frac{1}{L_s}\frac{1}{L_r}\frac{1}{L_o}\frac{1}{K_B} \tag{17}$$

where:

$G$ is Antenna gain of the receiver,
$T$ is System noise temperature,
$L_s$ is free space loss,
$L_r$ is rain attenuation loss,
$L_o$ is gaseous atmospheric loss,
$K_B$ is Boltzman's constant.

## 2.3 *Communication Subsystem Simulator GUI*

The subsystem is divided into input phase and output phase, the input phase divide into:

1. Location of the ground station: it consists of a small database of some positions for ground stations based on the latitude and longitude of the ground station.
2. Satellite type and its longitude.
3. General initial data: it consists of back off loss and losses disappointment both in db.

4. Ground station part: it consists of transmitter and receiver parameters.
5. Satellite station part: it consists of transmitter and receiver parameters.

While the output phase divided into:

1. Link Geometry: consist of Azimuth, Lifting angle in degrees and distance in KM
2. Upload link parameters
3. Downlink parameters
4. Total Link which determines the total carrier to noise ratio.

### 2.3.1 Graphical User Interface

In this window, we present the GUI of the satellite simulator as shown in Fig. 4.

### 2.3.2 GUI After Running the Simulator

After entering the inputs and run the simulator, the output (Link Geometry, Upload link and Down link) as shown in Fig. 5.

## *2.4 Data Packets*

We create look up tables for data in each mode as follow:

1. Each system is denoted by unique ID as shown in Table 1.

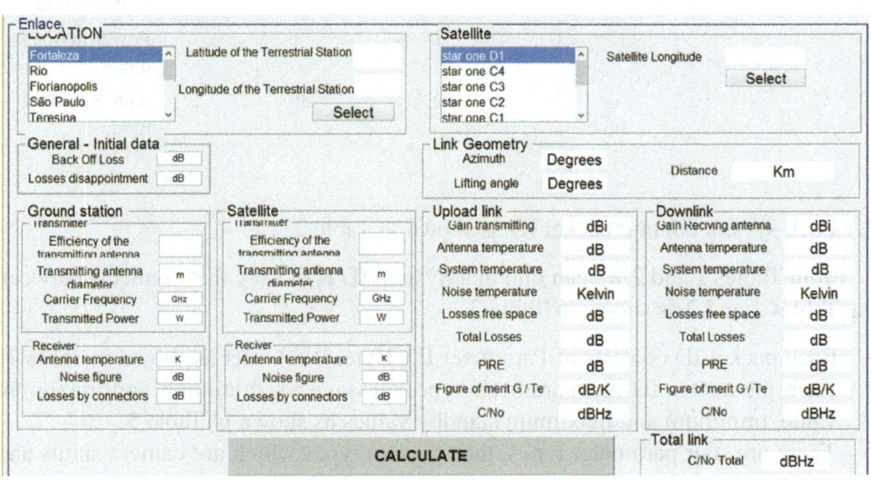


**Fig. 4** The GUI of the communication subsystem simulator

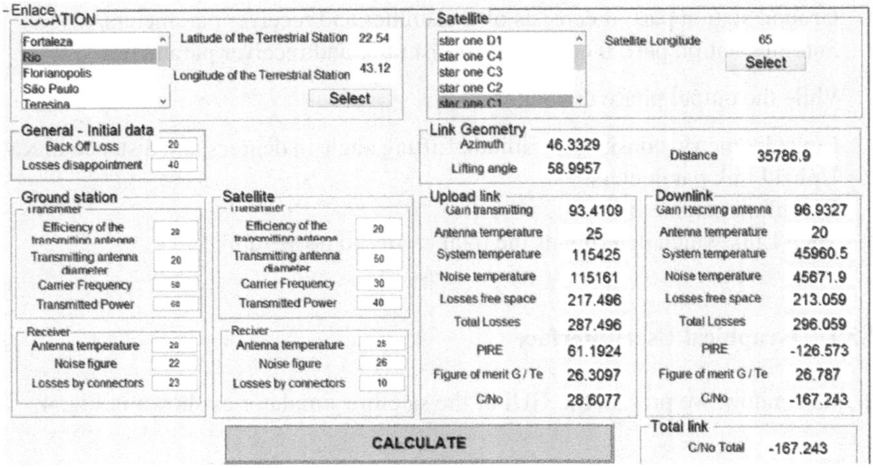


**Fig. 5** The output of the communication subsystem simulator after running

**Table 1** The four systems and their ID's

| System ID | Name |
|---|---|
| 1 | Power |
| 2 | ADCS |
| 3 | COMM |
| 4 | OBC |

**Table 2** Packet ID is assigned for each System ID

| Packet ID | System ID | Packet address | Packet destination |
|---|---|---|---|
| 11 | 1 | 5 | Power in standby mode |
| 21 | 1 | 6 | Power in image mode |
| 12 | 2 | 30 | ADCS normal mode |

2. Each system contain packet ID as shown in Table 2.

From Tables 1 and 2 we can find that System ID is the key that connects between the Tables 1 and 2 as shown in Table 3.

3. Each packet ID consists of Parameter ID, Byte, Bit, Order as shown in Table 4.
4. Each parameter ID has parameter type, message id, minimum and maximum value, minimum and maximum standby values as shown in Table 5.
5. There are four parameter types, the first two types which are camera status and command status hold texts, the third type which is power depends on given function and the fourth type is time as following Tables 6, 7, 8 and 9.

**Table 3** Concatenation between Tables 1 and 2

| Packet ID | System ID | Packet address | Packet destination | Name |
|---|---|---|---|---|
| 11 | 1 | 5 | Power in standby mode | Power |
| 21 | 1 | 6 | Power in image mode | Power |
| 12 | 2 | 30 | ADCS normal mode | ADCS |

**Table 4** Size and order for each packet

| Packet ID | Parameter ID | Byte | Bit | Order |
|---|---|---|---|---|
| 11 | 1 | 1 | 1 | 0 |
| 11 | 2 | 1 | 2 | 0 |
| 11 | 3 | 1 | 28 | 0 |
| 11 | 4 | 1 | 96 | 0 |
| 11 | 5 | 1 | 128 | 0 |
| 11 | 6 | 2 | 255 | 1 |
| 11 | 6 | 3 | 255 | 0 |
| 12 | 1 | 5 | 8 | 0 |

**Table 5** The ranges for each parameter in normal and standby mode

| Parameter ID | Parameter type | Message ID | Min value | Max value | Min standby | Max standby | PName |
|---|---|---|---|---|---|---|---|
| 1 | 1 | 1 | 0 | 1 | 0 | 1 | P1 |
| 2 | 11 | 2 | 0 | 1 | 0 | 1 | P2 |
| 3 | 2 | 1 | 0 | 1 | 0 | 1 | P3 |
| 4 | 0 | 0 | 0 | 0 | 0 | 0 | P4 |
| 5 | 10 | 0 | 0 | 0 | 0 | 0 | P5 |
| 6 | 3 | 1 | 0 | 3 | 1 | 1.5 | P6 |
| 7 | 4 | 1 | 1100 | 3.0125 e+05 | 1100 | 3.0125 e+05 | P7 |

**Table 6** Camera status description

| Parameter type | Message ID | Code | Description |
|---|---|---|---|
| 1 | 1 | 0 | Camera 1 S/W OFF |
| 1 | 1 | 1 | Camera 1 S/W On |
| 1 | 2 | 0 | Camera 2 S/W OFF |
| 1 | 2 | 1 | Camera 2 S/W On |

**Table 7** Command status description

| Parameter type | Message ID | Code | Description |
|---|---|---|---|
| 2 | 10 | 0 | No error |
| 2 | 10 | 1 | Command not explained |
| 2 | 10 | 2 | No response |

**Table 8** Power status description

| Parameter type | Message ID | Equation |
|---|---|---|
| 3 | 1 | 7.8125e−05 |

**Table 9** Time status description

| Parameter type | Message ID | Reference time |
|---|---|---|
| 4 | 1 | 1100 |

# 3 Satellite Communications System Segments

## *3.1 The Ground Segment (GS)*

The ground segment (GS) consists of the earth stations and other ground-based facilities used for communications traffic. With some systems, such as with the global positioning system (GPS), broadcasting satellite service (BSS) systems—also called direct broadcasting service (DBS), very small aperture terminal (VSAT) networks, and some military satellites, earth stations consist entirely of user terminals that interface directly with the space segment. In this case, the ground segment may be called the **user segment** [12].

## *3.2 The Space Segment (SS)*

The space segment (SS) comprises of at least one satellites in space, including both dynamic and extra satellites. A gathering of dynamic satellites is said to shape a group of stars. The dispatch vehicle and the majority of the offices required to dispatch satellites and place them in circle are likewise viewed as a component of the space portion [12].

### *3.3 The Control Segment (CS)*

The control segment (CS) incorporates the majority of the ground gear and offices that are required for task, control, checking and the executives of the space portion and, in numerous frameworks, the board of the earthbound system. Data is transmitted over free-space links. A one-way link from the ground to the satellite is called an **uplink**. A link from the satellite to the ground is a **downlink** [12] (Fig. 6).

## 4 Satellite Applications

The geostationary earth orbit (GEO) is in the equatorial plane at an altitude of 35,786 km with a period of one sidereal day (23 h 56 m 4.09 s). This orbit is sometimes called the Clarke orbit in honor of Arthur C. Clarke who first described its usefulness for communications in 1945. GEO satellites appear to be almost stationary from the ground (subject to small perturbations) and the earth antennas pointing to these satellites may need only limited or no tracking capability. An orbit for which

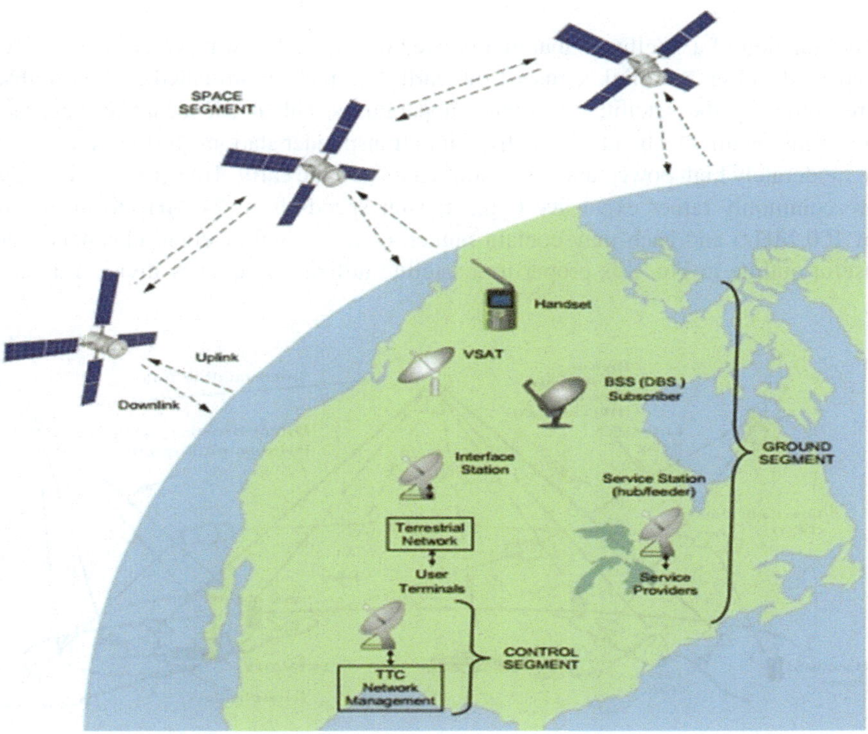


**Fig. 6** Satellite communications system segments

the highest altitude (apogee) is greater than GEO is sometimes referred to as high earth orbit (HEO). Low earth orbits (LEO) typically range from a few hundred km to about 2000 km. Medium earth orbits (MEO) are at intermediate altitudes. Circular MEO orbits, also called Intermediate Circular Orbits (ICO) have been proposed at an altitude of about 10,400 km for global personal communications at frequencies designated for Mobile Satellite Services (MSS) [6].

LEO systems for voice communications are called Big LEOs. Constellations of so-called Little LEOs operating below 1 GHz and having only limited capacity have been proposed for low data rate non-voice services, such as paging and store and forward data for remote location and monitoring, for example, for freight containers and remote vehicles and personnel [4].

Initially, satellites were used primarily for point-to-point traffic in the GEO fixed satellite service (FSS), e.g., for telephony across the oceans and for point-to-multipoint TV distribution to cable head end stations. Large earth station antennas with high-gain narrow beams and high uplink powers were needed to compensate for limited satellite power. Figure 7 depicts several kinds of satellite links and orbits.

## 5 Satellite Functions

The function of a satellite is that of a twisted pipe quasilinear repeater in space. As appeared in Fig. 8, uplink signals from earth terminals coordinated at the satellite are gotten by the satellite's reception apparatuses, enhanced, meant an alternate downlink recurrence band, channelized into transponder stations, further intensified to moderately high power, and retransmitted toward the earth. Transponder channels are commonly rather expansive (e.g., transfer speeds from 24 MHz to in excess of 100 MHz) and each may contain numerous individual or client channels. The useful outline in Fig. 8 is proper to a satellite utilizing frequency-division duplex

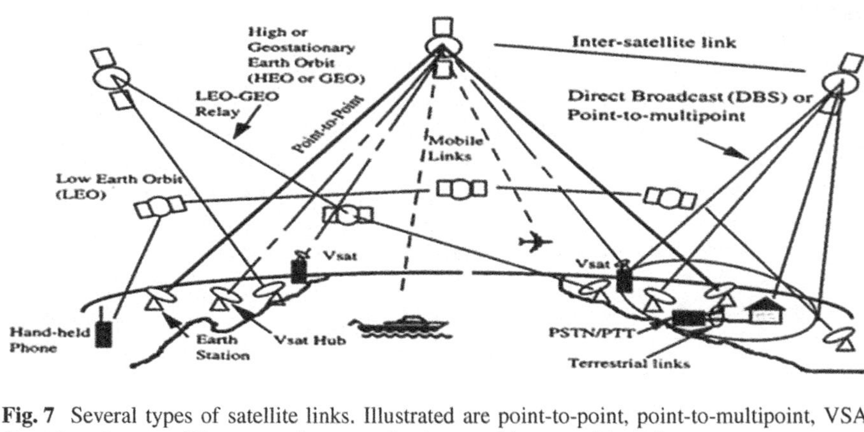


**Fig. 7** Several types of satellite links. Illustrated are point-to-point, point-to-multipoint, VSAT, direct broadcast, mobile, personal communications, and inter-satellite links

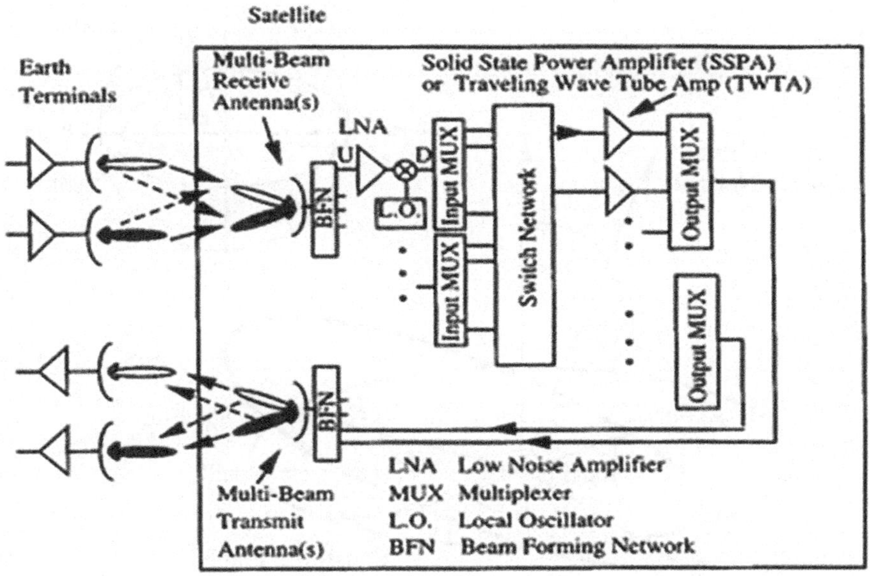


**Fig. 8** A satellite repeater receives uplink signals (U), translates them to a downlink frequency band (D), channelizes, amplifies to high power, and retransmits to earth. Multiple beams allow reuse of the available band. Interference (dashed lines) can limit performance. Down conversion may also occur after the input multiplexers. Several intermediate frequencies and down conversions may be used

(FDD), which refers to the fact that the satellites use separate frequency bands for the uplink and downlink and where both links operate simultaneously. This diagram also illustrates a particular multiple access technique, known as frequency-division multiple access (FDMA), which has been prevalent in mature satellite systems [9].

## 6 Satellite Orbits and Pointing Angles

Solid correspondence to and from a satellite requires a learning of its position and speed with respect to an area on the earth.

A satellite, having mass m, in circle around the earth, having mass Me, crosses a circular way to such an extent that the diffusive power because of its speeding up is adjusted by the world's gravitational fascination, prompting the condition of movement for two bodies:

$$\frac{d^2\mathbf{r}}{dt^2} + \frac{\mu}{r^3}\mathbf{r} = 0 \tag{18}$$

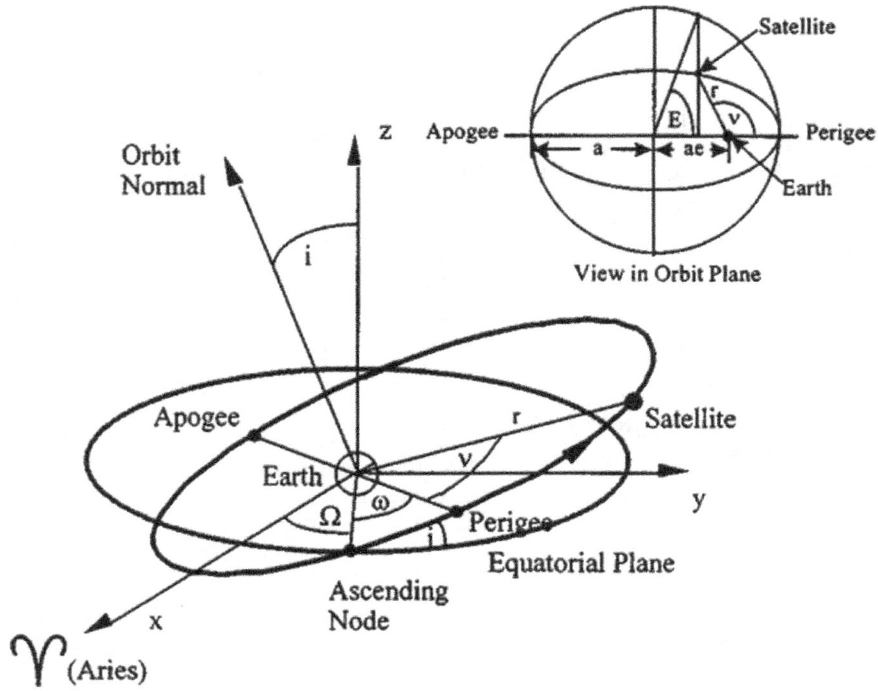


**Fig. 9** Orbital elements

where r is the radius vector joining the earth's center and the satellite and m = G (m + Me)ª GMe = 398,600.5 $km^3/s^2$ is the product of the gravitational constant and the mass of the earth. Since m ≪ Me, the focal point of revolution of the two bodies might be taken as the world's middle, which is at one of the central purposes of the circle oval [7].

Figure 9 delineates the orbital components for a geocentric right-gave organize framework where the x pivot focuses to the principal purpose of Aries, that is, the settled position against the stars where the sun's clear way around the earth crosses the world's tropical plane while going from the southern toward the northern side of the equator at the vernal equinox. The z pivot focuses toward the north and the y hub is in the tropical plane and indicates the winter solstice. The components demonstrated are longitude or right climb of the rising hub W estimated in the tropical plane, the circle's tendency edge I with respect to the central plane; the oval semimajor pivot length a, the oval capriciousness e, the contention (edge) of perigee w, estimated in the circle plane from the rising hub to the satellite's nearest way to deal with the earth; and the genuine irregularity (edge) in the circle plane from the perigee to the satellite n [3, 6, 8].

## 7 Satellite Links

Satellite links employ microwave frequencies above 1 GHz—the upper end is constrained to around 30 GHz for at present dynamic employments. The microwave designing procedure is the same as the training created amid and following World War II, when the utilization of this medium was quickened for radar and interchanges. While the standards continue as before, numerous advancements in computerized handling, microelectronics, programming, and exhibit receiving wires permit more alternatives for new applications. In this part, we quickly survey the rudiments of the satellite connection and relate it however much as could reasonably be expected to the necessities of the application. Different access frameworks, including frequency division multiple access (FDMA), time division multiple access (TDMA), and code division multiple access (CDMA), are also discussed and their strengths and weaknesses identified. When this audit is finished, we consider the famous recurrence groups utilized in business satellite correspondence (i.e., L, S, C, X, Ku, and Ka) alongside higher frequencies (Q- and V-groups), and additionally space-based optical interchanges [11].

### *7.1 The Basic Satellite Link*

Figure 9 shows satellite link in its simplest form, a satellite link carrying a duplex (two-way) communication circuit: the earth station A transmits to the satellite an uplink (U/L) carrier wave (modulated by the baseband signal, i.e. by the signal from the message source transmitted by the user terminal) at radio frequency (RF) $F_{u1}$ (e.g. 5980 MHz). The satellite antenna and transponder system receives this carrier and, after frequency conversion from $F_{u1}$ to $F_{d1}$ (e.g. 5980 MHz − 2225 MHz = 3755 MHz), amplifies and re-radiates it as a downlink (D/L) wave which is received by the earth station B. To establish the return link, B transmits a U/L carrier at another RF $F_{u2}$ (e.g. 6020 MHz) which is received by A at the converted D/L RF $F_{d2}$ (e.g. 6020 MHz − 2225 MHz = 3795 MHz) [11].

### *7.2 Design of the Satellite Link*

The satellite connection is likely the most essential in microwave interchanges since an observable pathway way regularly exists between the Earth and space. This implies a nonexistent line stretching out between the transmitting or accepting Earth station and the satellite reception apparatus goes just through the environment and not ground snags Such a connection is represented by free-space proliferation with just restricted variety regarding time because of different constituents of the climate. Free-space lessening is dictated by the reverse square law, which expresses that the power got is contrarily relative to the square of the separation. A similar law applies to the measure of light that achieves our eyes from a far off point source, for example, a vehicle front lamp or star. There are, nonetheless, some of extra impacts that create a lot of corruption and time variety. These include rain, terrain effects such as absorption by trees and walls, and some less-obvious impairment produced by unstable conditions of the air and ionosphere [4, 7].

It is the activity of the correspondence specialist to recognize the majority of the huge commitments to execution and ensure that they are appropriately considered. The required elements incorporate the execution of the satellite itself, the design and execution of the uplink and downlink Earth stations, and the effect of the engendering medium in the recurrence band of intrigue. Likewise essential is the productive exchange of client data over the applicable interfaces at the Earth stations, including such issues as the exact idea of this data, information convention, timing, and the media communications interface gauges that apply to the administration [4].

An appropriate building strategy ensures that the application will go into activity as arranged, meeting its goals for quality and dependability. The RF carrier in any microwave interchanges interface starts at the transmitting gadgets and engenders from the transmitting radio wire through the mechanism of free space and absorptive climate to the getting receiving device (antenna), where it is recouped by the accepting hardware. Like your vehicle FM radio or some other remote transmission, the transporter is regulated by a baseband flag that exchanges data for the specific application. The initial phase in structuring the microwave interface is to distinguish the general necessities and the basic segments that decide execution [6].

For this reason, the essential game plan of the connection appeared in Fig. 10. This model demonstrates an extensive center point kind of Earth station in the uplink and a little VSAT in the downlink; the satellite is spoken to by a basic recurrence deciphering sort of repeater (e.g., a bowed pipe). Most geostationary satellites utilize bowed pipe repeaters since these permit the largest scope of administrations and correspondence methods. Bidirectional (duplex) correspondence happens with a different transmission from each Earth station. Because of the simple idea of the radio recurrence connect, every component contributes a gain or misfortune to the connection and may include commotion and impedance [6, 8] (Fig. 11).

## 7.3 *Quantities for a Satellite RF Link*

Figure 12 illustrates the elements of the radio frequency (RF) link between a satellite and earth terminals. The overall link performance is determined by computing the link equation for the uplink and downlink separately and then combining the results along with interference and intermodulation effects. For a radio link with only thermal noise, the received carrier-to-noise power ratio is:

$$\left(\frac{c}{n}\right) = (p_t g_t)\left(\frac{1}{4\pi r_s^2}\right)\left(\frac{g_r}{T}\right)\left(\frac{1}{k}\right)\left(\frac{\lambda^2}{4\pi}\right)\left(\frac{1}{a}\right)(\rho)\left(\frac{1}{b}\right) \tag{19}$$

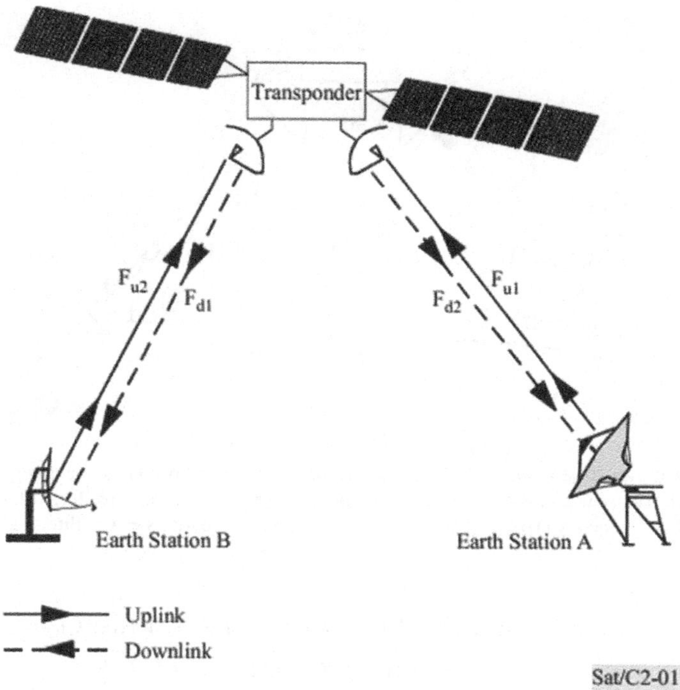


**Fig. 10** The basic satellite link

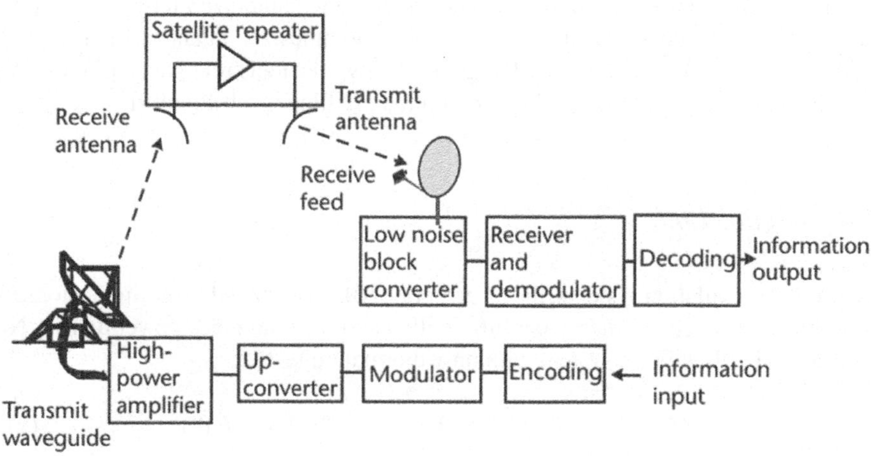


**Fig. 11** Critical elements of the satellite link

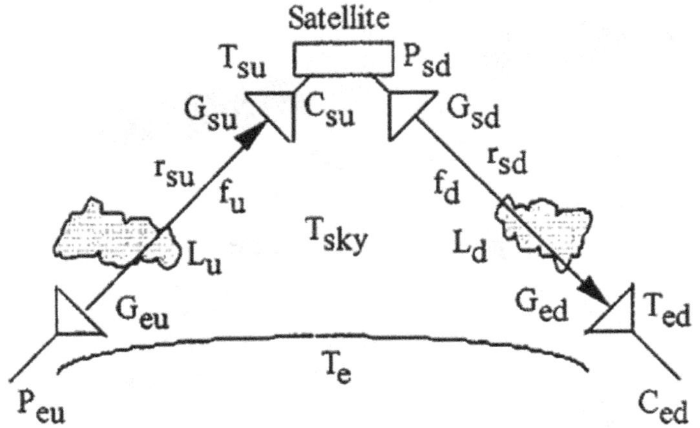


**Fig. 12** Quantities for a satellite RF link. **P** = transmit power (dBW). **G** = antenna gain (dBi). **C** = received carrier power (dBW). **T** = noise temperature (K). **L** = dissipative loss (dB). **rs** = slant range (m). **f** = frequency (Hz). **u** = uplink. **d** = downlink. **e** = earth. **s** = satellite

$$(C/N) = EIRP - 10\log\left(4\pi r_s^2\right) + (G_r - 10\log T) \\ + 228.6 - 10\log\left(4\pi/\lambda^2\right) - A + \Gamma - B \qquad (20)$$

where the subscripts in Eq. (18) refer to transmit (t) and receive (r). Lower case terms are the actual quantities in watts, meters, etc. and the capitalized terms in Eq. (19) correspond to the decibel (dB) versions of the parenthesized quantities in Eq. (18). For example, EIRP = P + G = 101logp + 101logg decibels relative to 1 W (dBW) and the expression (C/N) should be interpreted as 10logc − 10logn [9].

## 7.4 *Digital Links*

For digital modulation systems, the bit error rate (BER) is related to the dimensionless ratio (dB difference) of energy per bit, Eb dB J and the total noise power density No = 10log(kT) dB J. For a system with only thermal noise $N_o$.:

$$(E_b/N_o) = (C/N) + B - R = (C/N_o) - R\,\text{dB} \qquad (21)$$

where R = 10log (bit rate in bit/s), B is the bandwidth (dB Hz), and ($C/N_o$) is the carrier-to-thermal noise density ratio, that is, (C/N) normalized to unit bandwidth [6].

## 8 Satellite Communication Advantages

- **Global Availability**: Correspondences satellites cover all land masses and there is developing ability to serve oceanic and even aeronautical markets. Clients in country and remote locales around the globe who can't acquire fast Internet access from an earthbound supplier are progressively depending on satellite correspondences.
- **Superior Reliability**: Satellite correspondences can work autonomously from earthly framework. At the point when earthly blackouts happen from man-made and regular occasions, satellite associations stay operational.
- **Superior Performance**: Satellite is unmatched for communicated applications like TV. For two-way IP organizes, the speed, consistency and end-to-end control of the present propelled satellite arrangements are bringing about more noteworthy utilization of satellite by partnerships, governments and consumers.
- **Immediacy and Scalability**: Extra receive places, or hubs on a system, can promptly be included, once in a while inside hours. Everything necessary is ground-based hardware. Satellite has demonstrated its incentive as a supplier of "moment framework" for business, government and crisis help correspondences.
- **Versatility**: Satellites adequately bolster on a worldwide premise all types of interchanges running from straightforward purpose of-offer approval to data transmission serious media applications. Satellite arrangements are very adaptable and can work freely or as a component of a bigger system.

## 9 Satellite Communication Disadvantages

- Satellite production requires additional time. Besides, satellite plan and improvement requires greater expense.
- Satellite once propelled, requires to be observed and controlled on customary periods with the goal that it stays in the circle.
- Satellite has life which is around 12–15 years. Because of this reality, another dispatch must be arranged before it progresses toward becoming un-operational.
- Redundant parts are utilized in the system plan. This cause more expense in the establishment stage.
- In the instance of LEO/MEO, substantial numbers of satellites are expected to cover sweep of earth. In addition, satellite perceivability from earth is for brief span which requires quick satellite to satellite handover. This makes framework extremely mind boggling.

## 10 Conclusion

The design features, implementations, and validation of this SIM were displayed. The Proposed SIM can be utilized to approve rocket structure and measuring gauges by playing out a coordinated time reenactment of the shuttle. This distinguishes asset bottlenecks or deficiencies coming about because of disentangled suspicions. Since Proposed SIM is a time sensitive reproduction, discrete occasions and obligation cycles can be demonstrated and their subsequent effects can be assessed across all the spacecraft. Failure modes and operational possibilities can be assessed enabling the investigator to design activities (imagine a scenario in which situations) and enhance the rocket execution for a scope of mission situations. The Proposed SIM interface enables the examiner to effectively change framework utilitarian structures by means of square charts and to effortlessly refresh execution qualities of framework parts with parameter input menus. By changing explicit parameters in a model, the client can survey the effects of utilizing diverse advancements.

## References

1. P. Pathak, X. Feng, P. Hu, P. Mohapatra, Visible light communication, networking and sensing: a survey, potential and challenges. IEEE Commun. Surv. Tutor. **17**(4), 2047–2077 (2015) (fourth quarter)
2. NASA-AMES, Mars Climate Modeling Center, http://spacescience.arc.nasa.gov/mars-climate-modeling-group/brief.html, accessed online on 29 Mar 2016 [Online]. Available http://spacescience.arc.nasa.gov/mars-climate-modeling-group/brief.html
3. D. Amanor, W. Edmonson, F. Afghah, Presentation slides: utility of light emitting diodes for inter-satellite communication in multi-satellite networks, in *2016 IEEE International Conference on Wireless for Space and Extreme Environments*, Aachen (2016)
4. A. Alonso-Arroyo, V.U. Zavorotny, A. Camps, Sea ice detection using GNSS-R data from UK TDS-1, in *Proceedings of the 2016 IEEE International Geoscience Remote Sensing Symposium*, IEEE (2016), pp. 2001–2004
5. Space Studies Board, Achieving science with cubesats—thinking inside the box, National Academy of Sciences, Engineering and Medicine, Technical Report (2016) [Online]. Available https://www.nap.edu/catalog/23503/achieving-science-with-cubesats-thinking-inside-the-box
6. A. Alonso-Arroyo et al., On the correlation between GNSS-R reflectivity and L-band microwave radiometry. IEEE J. Sel. Top. Appl. Earth Observ. Remote Sens. **9**(12), 1–18 (2016)
7. Hyuk Park et al., A Generic level 1 simulator for spaceborne GNSS-R missions and application to GEROS-ISS ocean reflectometry, IEEE J. Sel. Top. Appl. Earth Observ Remote Sens. **10**(10), 4645–4659 (2017)
8. M. Unwin, P. Jales, J. Tye, C. Gommenginger, G. Foti, J. Rosello, Spaceborne GNSS-reflectometry on TechDemoSat-1: early mission operations and exploitation. IEEE J. Sel. Top. Appl. Earth Observ. Remote Sens. **9**(10), 4525–4539 (2016)
9. J. Wickert et al., GEROS-ISS: GNSS reflectometry radio occultation and scatterometry onboard the international space station. IEEE J. Sel. Top. Appl. Earth Observ. Remote Sens. **9**(10), 4552–4581 (2016)
10. H. Park, A. Camps, D. Pascual, Y. Kang, R. Onrubia, GARCA/GEROS-SIM M2 (Instrument to L1 module) web online simulation tool (2017)
11. Sarthak Singhal, Amit Kumar Singh, CPW-fed octagonal super-wideband fractal antenna with defected ground structure. IET Microw. Antennas Propag. **11**(3), 370–377 (2017)

12. D. Amanor, Visible light communication physical layer development for inter-satellite communication. Ph.D. dissertation, North Carolina A&T State University (2017)

# Part II
# Telemetry Data Analytics and Applications

# Crop Yield Estimation Using Decision Trees and Random Forest Machine Learning Algorithms on Data from Terra (EOS AM-1) & Aqua (EOS PM-1) Satellite Data

**Roheet Bhatnagar and Ganesh Borpatra Gohain**

**Abstract** Agriculture is one of the most important sectors of Indian Economy. Indian agricultural sector accounts for 18 % of India's gross domestic product (GDP) and provides employment to 50% of the countrys workforce. Estimation of Crop Yield during the cropping season plays an important role for planners and policy-makers for decision making. It is one of the critical parts as it depends on different factors weather, soil and crop Management. Weather plays one of the important role for plant growth and development, soil is important for plants to get nutrients while crop management is important for planning the planting time and application of different management practices for better crop yield. Different approaches had been reported for estimating crop growth and development based on statistical and mathematical based models. In this study the authors have tried decision tree and random forest based Machine Learning approaches to estimate crop yield. Decision Support System for Agro-technology Transfer (DSSAT) simulation model is used to estimate crop yield for the period from 1981 to 2025. The datasets from India Meteorological Department (IMD) from 1981 to 2016 and National Oceanic and Atmospheric Administration (NOAA) RCP4.5 climatic variables from 2017 to 2025 were used in the current study. We can provide a decision system that is able to learn from the input variable and predict the plant growth and development in real time. Our result indicates that there is (R2 = 0.67) and RMSE of 281 kg/ha from data predicted from random forest and Crop Yield Estimation. The study makes use of MODIS data from Earth Observing System satellites namely Terra (EOS AM-1) & Aqua (EOS PM-1) from NASA. The study validates the predicted yield by comparing their values with

R. Bhatnagar (✉)
Department of Computer Science & Engineering, Manipal University Jaipur, Jaipur, India
e-mail: roheet.bhatnagar@jaipur.manipal.edu

G. B. Gohain
Munich RE, Mumbai, India
e-mail: GGohain@munichre.com



A. E. Hassanien et al. (eds.), *Machine Learning and Data Mining in Aerospace Technology*, Studies in Computational Intelligence 836,
https://doi.org/10.1007/978-3-030-20212-5_6

NDVI values. High NDVI values means more vegetation and more yield. Hence, it can be said that decision trees and random forests can be used in forecasting the crop yield.

## 1 Introduction

In-season crop yield estimation is important for planners and policymakers to take a decision, but timely and accurate crop yield forecast is important. A lot of research has been going on for estimation of crop yield using different approaches. Statistical Model, Remote sensing, and Crop simulation model being some of them. However still as on date, there are no such approach which can lead to an accurate prediction of crop yield estimation. With the advancement of computer technology, machine learning has been extensively used in different sectors. Machine learning can be used in Agriculture sector for estimation of crop yield and our study focuses on some of the ML based methods. Advancement of research, development and technology transfer the green revolution in worldwide has increased the crop production in 2009, By 2050 the agricultural production would need to increase by 70% due to emerging population that is expected to exceed by 9 billion. Advancement of Technology and data availability has added to big data concept in different sectors. Agriculture is one of the important sector where data plays an important role for decision making. Machine learning can be implemented in Agriculture data to get important data patterns, clusters, classification, segmentation and Predictions. With the increasing amount of data volumes, data scientist plays a major role in handling and processing Big data for meaningful but hidden information. The government has also adopted the technology of Big Data to make Smart Cities.

## 2 Related Work

The benefit of modern data mining methods which offer over current, time-honored method like stepwise regression modelling have been proven and cited by researchers working in the domain [6]. Literature also have references where the researchers have used a random forest modelling technique to study the climatic impact on sugarcane productivity in the Victoria, Bundaberg and condoning sugar mill regions in Australia [2]. From our literary reviews we understand that in order to increase the robustness and precision of prediction we can use multiple efforts for a response instead of using a single data set or model [2, 7]. Random forest should not be confused with the single decision tree [5], and there are many studies where Random Forest have been outperformed by traditional linear approaches [4, 9] because of different reasons e.g. the nature of the dataset etc. Random forest can be used in agricultural related applications e.g. studies have been carried out to predict mangos yield random forest has been used [8].

Random Forest is used in Big data analysis to investigate Nitrous Oxide ($N_2O$) emission [12], Leaf Nitrogen Level [1] and drought Forecasting [3]. Building a Predictive model for Sugarcane for leaf nitrogen levels from hyperspectral satellite images using Random forest regression [1], researchers also used the random forest regression to identify the most important predictor variables of $N_2O$ emission [12]. In the current study the Random forest is used to investigate the impact of climate characteristics in Rice crop productivity in Hisar District of Haryana a Northern State of India. The main advantage of using Random forest technique is that the association between predictors and the response variable used the ensemble learning approaches with the nonlinear and hierarchical relationship. Different Ensemble methods are being involved in making multiple attempts using different data or models for crop yield estimation. We can increase the accuracy of the response variable by using multiple efforts instead of using a single dataset or model. Researchers across the world use different approaches to predict the crop yield of which Random forest have been used extensively in different agriculture-related applications.

## 2.1 *Machine Learning—A Brief Overview*

These days the technology world is abuzz with terms like Machine Learning, Big Data Analytics, Deep Learning etc. but then what is Machine Learning? Well, simply put Machine Learning is a term associated with computer program which can learn and adapt itself to new situations and environments without human interference. According to Tom Mitchell, Carnegie Mellon University, Machine Learning is A computer program is said to learn from experience E with respect to some task T and some performance measure P, if its performance on T, as measured by P, improves with experience E.

It is a field of Artificial Intelligence and has gained popularity because of the numerous real world applications and problems that it can solve. Every sector and domain in today's world is generating lot of data, through various data sources and these data need to be processed to get hidden information and insights for the benefit of businesses. The companies and governments are aware of the new information gained through sniffing through Big Data using Machine Learning algorithms and methods. Machine Learning has found its applications in almost all the sectors e.g. financial, e-commerce, stock market, asset management etc.

Machine Learning algorithms and methods are broadly classified into two broad categories as:

- Supervised machine learning: The program is trained on a pre-defined set of training examples, which then facilitate its ability to reach an accurate conclusion when given new data.
- Unsupervised machine learning: The program is given a bunch of data and must find patterns and relationships therein.

Linear Regression, Logistic Regression, Decision Tree, SVM (Support vector machines), Naive Bayes, KNN (K nearest neighbors), K-Means, Random Forest, etc. are some of the commonly used Machine Learning algorithms. In this chapter the authors have applied Machine Learning algorithms namely Decision Trees and Random Forest in estimating the Crop Yield and the approach is discussed in detail in the subsequent sections.

## 2.2 *Machine Learning in Agriculture*

Estimation of crop yield during the cropping period play an important role. Machine learning is an emerging technology for agriculture and many researchers are working in this domain to come up with ML based methods for better and more accurate prediction of crop yield. With the use of machine learning, the agricultural sector too intends to improve technologically for increased crop production by better crop management by applying analytics on agricultural big data and time-series data. Crop growth stress detection can be done through machine learning using satellite image classification. The predicting capability for crop yield estimation with machine learning method can be improved and achieved by defining rules and looking for patterns in large datasets. Machine learning helps to self-improvise the predictive model. Decision tree and Random forest are two algorithms that are most popular and are extensively used for the real-time application. Machine learning can be used to develop a probability model and can consider all variables and can predict a certain outcome.

Machine learning can be used in technology development making it more accurate and precise by bringing in improvement in the existing processes. Machine learning can be used for plant breeding which helps us to understand as to how to genetically develop the crop variety.

## 2.3 *Crop Yield Estimation*

Accurate, timely and early estimation of crop yield during the cropping season plays a major impact and it impacts government policies. Crop yield estimation is required for different purposes. It can be used for Crop insurance, Delivery estimation, Planning harvesting and for storage the crop production and for cash flow. Crop yield estimation is required for monitoring the crop growth and development during the cropping period. Crop yield not only estimated yield it also gives out output as Leaf Area Index, Harvest Index, Maturity date, Biomass as well as about different crop stress factors such as Nitrogen stress and water stress.

# 3 Methodology Adopted

This section discusses the detailed methodology adopted in the current study which presents an approach different from traditional techniques for estimating the crop yield. This study has used both the crop simulation model and machine learning method based on decision tree and random forest, to predict the crop yield. Crop simulation model DSSAT is used to simulate crop yield from 1981 to 2025 with climatic data from IMD and NOAA containing crop management information, soil information and crop variety informations. After simulation the authors have used the ML methods/algorithms to predict the crop yield based on predictive and response variables. The decision tree has been utilized to build the tree structure and to predict the yield, based on the decision tree generated with different key variables. Random Forest approach has been applied for classification and regression. This method helps us to fit the training sample data in the subsequently used fitting regression model, to predict the crop yield with the remaining test data. Estimation of crop yields within the growing season is critical for making agricultural and food security decisions.

The authors have implemented and applied Machine Learning through a decision tree and random forest. To estimate crop yield we have considered various factors namely—the climatic variability, crop management information and crop phenology parameters. IMD weather data and NOAA RCP 4.5 climatic data are used. These climatic data sets are used as input to crop simulation model which is simulated to estimate the crop yield and other phenology parameters. Implementation of Random Forest and Decision trees on the estimated crop yield are carried out for better prediction. Random forest and Decision tree takes all the climatic variables and crop phenology information.

We have considered 11 predictor variables and 1 response variable. The predictors variables are year (YEAR), Planting data (DATE), Harvest Index at maturity (HIAM), Irrigation Amount (IRCM), Precipitation amount (PRCM), Leaf Area Index (LAIX), Maximum Temperature (TMAX), Minimum Temperature (TMIN), Solar Radiation (SRAD), and Day length period (DAYLA). These predictors variables were used in the random forest for better models predictions. These are the important factors for plant growth and development.

## *3.1 About DSSAT v4.6 Crop Simulation Model*

The key components required for crop growth simulation model is good quality weather, experimental and soil data. Data for simulation are in terms of process-related constants such as photosynthetic efficiency, data on partitioning of assimilates, and phonological development, and other external driving variables. The required input data for crop simulation model are agrometeorological data such as radiation and temperature, soil data describing the hydraulic properties which is required for soil-water balance, and crop data factors describing the physiological and morphological processes that govern crop growth. When these information requirements are not available, they may be estimated from existing databases and expert knowledgebase available with concerned government agencies, but they may not result in accurate yield estimation. Crop growth simulation modelling is useful tool to describe continuous crop growth and to estimate crop yield using environmental inputs, DSSAT (Decision Support System for Agro-Technology Transfer) is used extensively in addition to the other models such as WOFOST, Info Crop etc. Location and crop specific models are validated using data generated at respective places.

DSSAT is a decision support system that is designed to aid farmers in developing long term crop rotational strategies. Fifteen crop stimulation models (CERES: wheat, maize, rice, sorghum, millet, barely, sunflower, sugarcane, chickpea, tomato and pasture; SOYGRO, PNUTGRO, BEANGRO, SUBSTOR-potato) are accessible in DSSAT [10]. The crop models are developed to assess the influence of weather and management practices (cultivar selection, sowing time, plant population, initial condition, irrigation water, nitrogen schedule, mulching etc.) on crop growth and development on daily basis. Significant feature of DSSAT is the development of standards for data collection and formats for data acquisition and exchange. This allows any crop model of the family to share and access common soils and weather data. The models include: CERES-Barley, CERES-Maize, CERES-Millet, CERES-Rice, CERES-Sorghum and CERES-Wheat. Applications of system approach using comprehensive models like DSSAT to represent the total agricultural production systems require datasets on the different components, namely: (a) crop; (b) weather; (c) soil; and (d) management. Models require cultivar specific genetic coefficients for better prediction. These coefficients vary among the crop varieties in response to weather, soil and management practices. Estimation of genetic coefficients, hence, is vital in crop yield forecasting models. Field experiments will be conducted for different cultivars and observations will be made to estimate genetic coefficients [11].

### 3.1.1 Model Inputs

1. Crop data/cultivar file: Crop datasets include the genetic coefficients with genetic parameters that characterize the physiological and morphological processes determining crop growth, development and yield. Changes of genetic coefficient changes the overall characterize of the plant development. Crop cultivar consider for the forecast are dominant varieties grown by the farmers of the region. Water and nitrogen management parameters considered in the model were as per agronomical recommendation widely.
2. Genetic coefficient: Crop datasets include the genetic coefficients with genetic parameters that characterize the physiological and morphological processes determining crop growth, development and yield. Changes of genetic coefficient changes the overall characterize of the plant development.

3. Weather data: Weather data plays important role for yield estimation in crop simulation model because the model responds to variability in weather parameters. To simulate the crop growth and development weather parameters needed are daily values of maximum and minimum temperature, Bright sunshine hours (BSSH) or solar radiation, since light and temperature are one of the key parameters driving variables of plant processes.

### 3.2 Study Methodology Flow chart

We have used crop simulation model as well as Decision tree and Random forest algorithms. For every crop simulation model data is one of the important factor to start the simulation. Crop simulation model require Weather data, soil physicochemical properties, crop management details and crop genetic co-efficient. All these parameters are input variables to crop model. After the model gas been successfully simulated. We use the required information as input to Decision tree and random forest algorithms for finding the relationship between the simulation model and prediction variables. If the relationship between the simulated and the predicted value are positive we can further use the random forest and decision tree to estimate crop yield. Further Decision tree and Random forest can be used as a reference tools with crop simulation model.

### 3.3 Dataset Used

For our case study, we have considered IMD Gridded Weather data and NOAA Climate scenario RCP 4.5 data as given in Annexure 1 and Annexure 2 at the end of the chapter. For the crop simulation model, we have used the weather data, Crop Management information which includes the crop sowing data, fertilizer application, Irrigation applied and Initial soil condition. Soil information includes the soil hydro physical properties. After the model has been simulated successfully, we extracted the required information from the summary file from the crop simulation model. This information is further used as input to Decision tree and Random forest for prediction (Fig. 1).

The important features generated from a crop simulation model is used further as an input to the random forest and decision tree algorithms and the parameters include are as mentioned in Table 1.

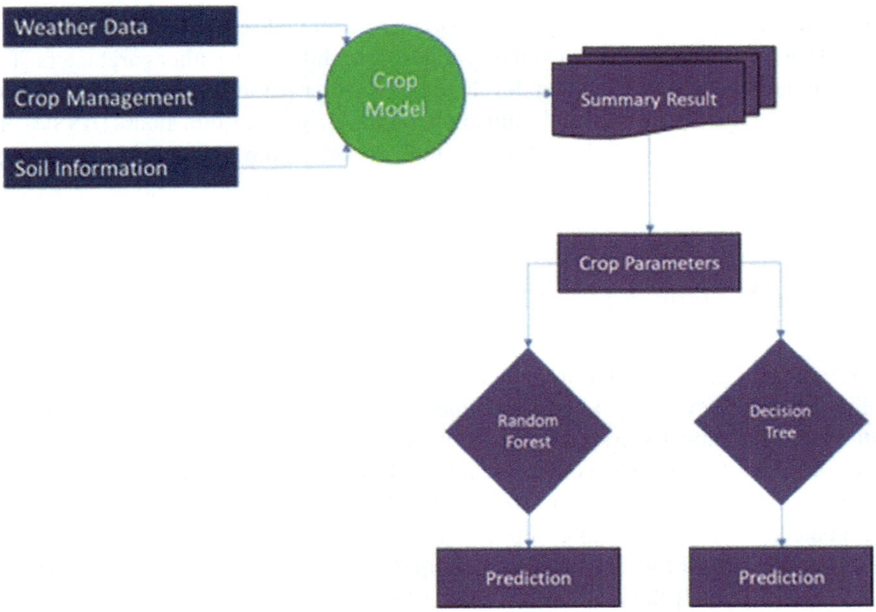


**Fig. 1** Flow chart showing the implementation of crop simulation modelling, decision tree and random forest

**Table 1** Table showing the different crop parameters generated from crop simulation model and implementing these datasets in decision trees and random forest inputs parameters

| | |
|---|---|
| CWAM | Tops wt kg/ha Tops weight at maturity (kg [dm]/ha) |
| HWAH | Harvested yield harvested yield (kg [dm]/ha) |
| BWAM | Byproduct kg/ha by-product produced (stalk) at maturity (kg[dm]/ha) |
| HIAM | Harvest index harvest index at maturity |
| LAIH | LAI harvest leaf area index, at harvest |
| IRCM | Irrig mm season irrigation (mm) |
| TMINA | Minimum temp C avg minimum air temperature (C) |
| TMAXA | Maximum temp C avg maximum air temperature (C) |
| SRADA | Avg solar rad average solar radiation ($MJ/m^2/d$), planting-harvest |
| DAYLA | Avg day (h) average daylength (h/d), planting to harvest |
| PRCP | Precip, plant total season precipitation (mm), planting to harvest |

## 4 Results and Discussions

### *4.1 Decision Tree Interpretation*

Decision tree iteratively splits the datasets into distinct subsets in a greedy fashion way. The regression tree minimize either by the MAE (Mean absolute error) or the MSE (Mean square error) within all the subsets variables. The resulting subsets with classification trees is splits and minimize entropy or Gini impurity. To predict the crop yield based on number of variables such as TMAX (Maximum Temperature), TMIN (Minimum Temperature), LAIX (Leaf Area Index at Maturity) and PRCM (Precipitation). Maximum depth of tree is kept at limiting to 3 levels. To predict the crop yield, a decision tree will traverse down the tree until it reach the leaf node. In each steps of the decision tree it splits the current subset into two The contribution of variable for a specific split that determined the split is define as the change in mean crop yield. For example if we consider the TMAX value = 32.0 it will fall in the leftmost leaf and it will predict as yield y = 1643.2 kg/ha. The tree structure is formed by the Decision tree which builds classification or regression models. The datasets are discredited into smaller and smaller subsets while parallelly it is associated with decision trees incrementally developed. The result formed is a tree with decision nodes and leaf nodes. The *p*-value method is associated with the hypothesis test. We have created two hypothesis one is the null hypothesis and the other is the alternative hypothesis. The null hypothesis is to be tested. If somehow the null hypothesis is not accepted the alternative hypothesis is considered acceptable.

### *4.2 Random Forest Interpretation*

Random forest takes mean contribution for a variable across all trees in the forest to determine the contribution of features. Random forest are integrally random, For crop yield different climatic conditions, soil hydro physical characteristics and crop management information are correlate to each other.

This result shows that there is positive trend and increasing relationship between crop yield and maximum temperature (TMAX). We can find the relationship between different parameters which influence the crop yield.

With Random forest we have predicted the crop yield with major influencing parameters. We have found that there is a good R2 value between simulated and predicted values. The predicted values is from Random forest algorithms and the simulated value is from Crop simulation model. The prediction from Random forest algorithms was quite good with the 67% of the variance in the response variable can be explained by the explanatory variable. The RMSE value was calculated as 281 kg/ha for rice crop. The yield simulated from crop simulated model and the yield predicted from the random forest shows the good performance of the algorithms. We divided the entire data into training and the test dataset where the probability for training dataset is 0.7 and for the test, the dataset is 0.3. We fitted the model with the training data set and predicted the result with the test dataset. We kept crop yield as the response variable and remaining datasets as the predictor variable. The predicted value for 2018 if the sowing data of the crop is 25 July is expected to be 1985 kg/ha whereas the simulated value is 1986 kg/ha with deviation of only 0.3% where the predicted value for 2019, if the sowing date of the crop is 25 July, is 2124 kg/ha as compared to the simulated to 2022 kg/ha with deviation of −5% from the simulated. This shows that a random forest is a good approach for estimating crop yield (Fig. 2).

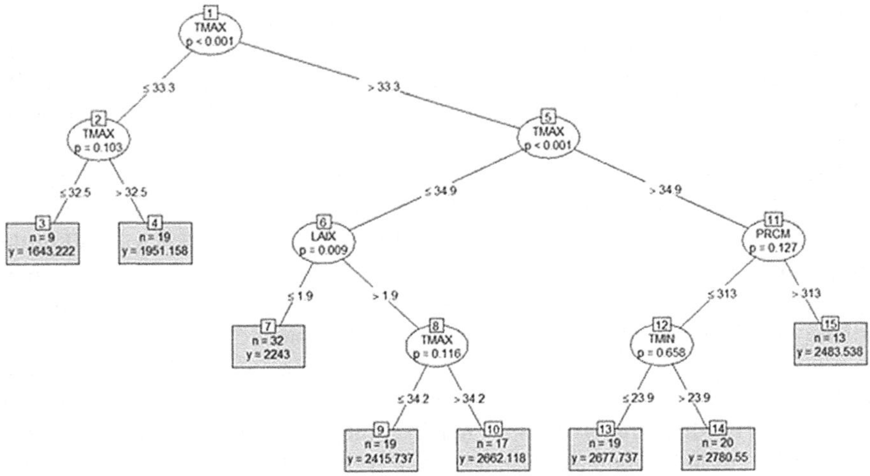


**Fig. 2** Result from decision tree where we have considered TMAX, TMIN, LAIX, PRCM for predicting the yield

## *4.3 Normalized Difference Vegetation Index as a Performance Measure*

Additionally the authors have made use of Normalized Difference Vegetation Index (NDVI) to corroborate the efficiency of ML algorithms in their study. NDVI is used as a measure to quantify vegetation by measuring the difference between near-infrared (which vegetation strongly reflects) and red light (which vegetation absorbs) and it's value ranges between $-1$ and $+1$. So, when we have negative values, its highly likely that its water. On the other hand, if you have a NDVI value close to $+1$, theres a high possibility that its dense green leaves. But when NDVI is close to zero, there isnt green leaves and it could even be an urbanized area (Figs. 3 and 4).

In order to establish the results of ML models we have compared the NDVI values with the predicted yield values as obtained using ML algorithms for sample years 2013, 2015 and 2016. Similarly, the comparison can be done for all other years as well. Due to the high temporal resolution, to calculate NDVI we have chosen MODIS as the optical sensor from MOD13Q1/MYD13Q1 (16 days VI composites at 250 m spatial resolution) version 6 products acquired by Terra and Aqua, respectively. For the current study, we have just calculated NDVI values for 2013, 2015 and 2016 for 24th October of each year. The NDVI value as calculated are 0.439, 0.414 and 0.435 for the years 2013, 2015 and 2016 respectively.

NDVI is calculated using the formula where NIR is Near Infrared band and IR is the Infrared band

$$NDVI = (NIR - R)/(NIR + R) \tag{1}$$

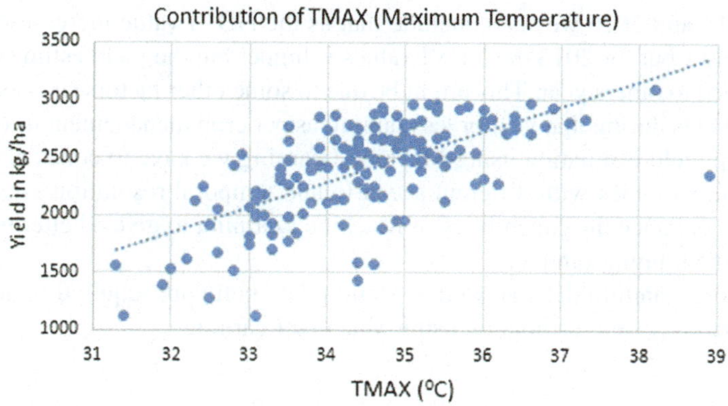


**Fig. 3** Figure showing the correlation between crop yield and maximum temperature (TMAX)

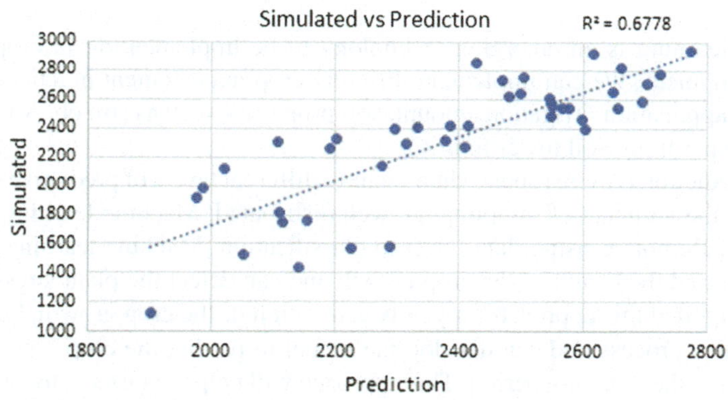


**Fig. 4** Figure showing the result predicted from random forest and result generated from crop simulation model

**Table 2** Predicted yield versus NDVI for sample years

| Year | Predicted yield | NDVI |
|---|---|---|
| 2013 | 2110 | 0.439 |
| 2015 | 2547 | 0.414 |
| 2016 | 2600 | 0.435 |

Using Machine Learning also we have calculated the predicted yield values for 2013, 2015 and 2016. Table 2 is for Yield from the Predictive model and NDVI from MODIS data.

Here for 2013, the yield is 2110 kg/ha and the NDVI value is 0.439, for 2015 the yield is 2547 kg/ha and the NDVI value is 0.414 and for 2016 the yield is 2600 kg/ha and the NDVI value is 0.435.

For 2015 and 2016 we can conclude that as the NDVI value increases the yield also increases but for 2013 the NDVI values is higher but the yield estimated is less as compared to others year. This might be due to some other factors such as extreme weather events during that year or less rainfall as per crop management information, so for crop yield estimation using Machine Learning we have to consider different sets of data variables with different scenario and temporal resolution satellite data which will enhance the capabilities of Machine Learning to give a better prediction based on the training sample.

Therefore, satellite data, as well as station data both, are required in arriving at an accurate crop yield estimation using Machine Learning.

## 5 Future Scope

Machine learning is an emerging technology to be implementing in crop growth and development. We can implement different crop management practices such as Fertilizer application, irrigation amount, soil properties such as soil ph., soil organic carbon, upper limit, soil lower limit.

The prediction is more robust when we have different levers of predictive variables and when the numbers of sampling are well distributed. Machine learning can also be used in Remote sensing data for crop classification. Machine learning can also be used to find the health of the crop growth and can detect the plant stress factors. We have applied this approach for yearly estimation of the crop growth, but we can partition this process and can develop the model to predict the crop yield at every stated during the cropping period. This approach will help the farmers to monitor the crop growth and can implement Fertilizer and Irrigation when required the planners and policymakers as well to monitor the crop growth and development.

## 6 Conclusion

The approach for this paper is to find the use of machine learning to predict crop yield and which we have applied for Rice Crop for Hisar district in India. Predicting the crop yield in advance helps the farmers to decide about the market and profits. It also helps in the crop insurance market for risk assessment. The sustainable solution for policy makers and planners helps in the improvement of economic environmental growth are important.

The random forest model was good in prediction crop yield for in the season crop yield estimates. The random forest can also be used for predicting the crop yield with the future climatic scenario data. This paper approach was to find the utilization of Crop simulation model as well as the implementation of Random forest in agriculture. Crop simulation models can be used for crop yield estimation. Crop simulation model required different sets of variables to predict the crop yield which takes input variables as weather parameters (Maximum and Minimum Temperature), Crop Management Information, soil information and crop variety. These inputs parameters make the crop sensitivity and response to the input variable. Weather parameters play one of the major roles in the sensitivity of the model. As a crop simulation model required different numbers of input variables. These input datasets can be used in the Random forest to develop our predictive models. Once the predictive model has been developed we can implement for estimating crop yield.

Our approach was to estimate the crop yield with 11 predicted variables and 1 response variable using the random forest. The random forest can also be used for different agricultural applications. It can be used for predicting different crop parameters which include crop biomass, leaf area index, Harvested Index, crop maturity period, crop yield etc. as well as input parameters can also be predicted for crops which will include the application of Irrigation and Fertilizer, sowing date based on the weather parameters. Important of Machine learning is that the prediction is much reliable when there is more training dataset. It helps to fit the regression model with different factors with different combinations and build a better predictive model. This approach using of machine learning is important and support the use of Machine Learning with Random forest and decision trees for Agricultural practices and to utilize for different crops and locations for estimating crop yield during the cropping period.

**Acknowledgements** The authors would like to thank to IMD to provide the IMD Gridded data for one location and NOAA for RCP 4.5 data.

## Appendix 1

Input Dataset in Decision Tree and Random forest. Weather data (PRCM, TMAX, TMIN, SRAD, DAYLA) are Climatic information whereas (DATE, CWAM, HWAM, LAIX, IRCM are from crop simulation model).

| Year | Date | CWAM | HWAM | HIAM | LAIX | IRCM | PRCM | TMAX | TMIN | SRAD | DAYLA |
|---|---|---|---|---|---|---|---|---|---|---|---|
| 1981 | 177 | 6171 | 2127 | 0.345 | 1.6 | 1451 | 351 | 35.5 | 25.4 | 18.6 | 13 |
| 1982 | 177 | 8039 | 2715 | 0.338 | 1.7 | 1484 | 211 | 36.6 | 24.7 | 20.2 | 13 |
| 1983 | 177 | 6809 | 2384 | 0.35 | 1.6 | 1440 | 366 | 34.8 | 25.3 | 18.1 | 13 |
| 1984 | 177 | 6390 | 2211 | 0.346 | 2 | 1673 | 321 | 33.1 | 23.7 | 17.5 | 12.9 |
| 1985 | 177 | 7071 | 2435 | 0.344 | 2.1 | 1434 | 352 | 35.5 | 25 | 19 | 13 |
| 1986 | 177 | 7797 | 2794 | 0.358 | 1.8 | 1501 | 313 | 35.1 | 23.2 | 20.1 | 12.9 |
| 1987 | 177 | 6790 | 2353 | 0.347 | 1.1 | 1796 | 50 | 38.9 | 26 | 20.8 | 12.9 |
| 1988 | 177 | 7992 | 2749 | 0.344 | 2.3 | 1390 | 611 | 34.4 | 25.3 | 17.6 | 13 |
| 1989 | 177 | 7799 | 2864 | 0.367 | 1.6 | 1501 | 131 | 36.4 | 24.2 | 20.4 | 12.9 |
| 1990 | 177 | 6163 | 2165 | 0.351 | 1.9 | 1386 | 431 | 34.6 | 24.9 | 18.3 | 12.9 |
| 1991 | 177 | 8289 | 2983 | 0.36 | 1.6 | 1430 | 228 | 36.6 | 24.7 | 20.2 | 12.9 |
| 1992 | 177 | 7520 | 2558 | 0.34 | 1.9 | 1676 | 357 | 34.3 | 23.1 | 19.3 | 12.9 |
| 1993 | 177 | 5214 | 1746 | 0.335 | 1.7 | 1645 | 658 | 33 | 24.6 | 16.8 | 12.9 |
| 1994 | 177 | 5245 | 1676 | 0.319 | 1.9 | 1554 | 828 | 32.6 | 23.8 | 16.7 | 12.9 |
| 1995 | 177 | 6662 | 2236 | 0.336 | 1.6 | 1351 | 403 | 35.1 | 25.3 | 18.1 | 12.9 |
| 1996 | 177 | 7760 | 2823 | 0.364 | 2.1 | 1355 | 297 | 35.2 | 24.2 | 19.3 | 12.9 |
| 1997 | 177 | 7525 | 2637 | 0.35 | 1.7 | 1329 | 455 | 35.2 | 24 | 19.8 | 12.9 |
| 1998 | 177 | 6539 | 2321 | 0.355 | 1.7 | 1443 | 356 | 36 | 26.1 | 18.6 | 13 |
| 1999 | 177 | 8310 | 2941 | 0.354 | 1.9 | 1454 | 151 | 36.9 | 25.2 | 20.3 | 13 |
| 2000 | 177 | 7931 | 2751 | 0.347 | 1.9 | 1537 | 75 | 36.4 | 23.8 | 20.7 | 12.9 |
| 2001 | 177 | 7074 | 2475 | 0.35 | 2.1 | 1435 | 387 | 35.1 | 23.2 | 20.2 | 12.9 |
| 2002 | 177 | 8061 | 2806 | 0.348 | 1.5 | 1527 | 81 | 37.1 | 25.4 | 20 | 12.9 |
| 2003 | 177 | 6635 | 2412 | 0.364 | 2.4 | 1336 | 432 | 34.2 | 24.5 | 18.1 | 12.9 |
| 2004 | 177 | 8390 | 2777 | 0.331 | 1.7 | 1455 | 184 | 36 | 24.8 | 19.8 | 13 |
| 2005 | 177 | 6161 | 2249 | 0.365 | 1.5 | 1440 | 458 | 34.6 | 24.8 | 18.3 | 12.9 |
| 2006 | 177 | 7852 | 2771 | 0.353 | 2.1 | 1483 | 220 | 35 | 24.7 | 18.8 | 12.9 |
| 2007 | 177 | 8157 | 2904 | 0.356 | 1.8 | 1442 | 187 | 35.4 | 25.4 | 18.8 | 13 |
| 2008 | 177 | 7338 | 2492 | 0.34 | 2.2 | 1421 | 387 | 34.6 | 24.8 | 18.4 | 12.9 |
| 2009 | 177 | 8066 | 2787 | 0.345 | 1.8 | 1486 | 351 | 36.8 | 25.4 | 20 | 13 |
| 2010 | 177 | 7415 | 2580 | 0.348 | 2.3 | 1164 | 660 | 34.2 | 25.3 | 17.4 | 12.9 |
| 2011 | 177 | 7409 | 2547 | 0.344 | 2 | 1365 | 341 | 34.5 | 24.9 | 18.1 | 12.9 |
| 2012 | 177 | 7434 | 2666 | 0.359 | 1.7 | 1393 | 393 | 35.4 | 26 | 18.1 | 13 |
| 2013 | 177 | 5697 | 1957 | 0.344 | 1.7 | 1407 | 588 | 34.9 | 25.7 | 17.9 | 13 |
| 2014 | 177 | 8108 | 2705 | 0.334 | 1.8 | 1469 | 203 | 36.3 | 25.9 | 19 | 13 |
| 2015 | 177 | 7618 | 2662 | 0.349 | 2.3 | 1427 | 300 | 35.1 | 24.8 | 18.6 | 13 |
| 2016 | 177 | 5773 | 1952 | 0.338 | 1.6 | 1431 | 341 | 35 | 25.5 | 18.1 | 13 |
| 2017 | 177 | 6284 | 1985 | 0.316 | 1.8 | 1644 | 499 | 33.5 | 22.9 | 24 | 12.9 |
| 2018 | 177 | 7456 | 2539 | 0.341 | 1.6 | 1443 | 310 | 35.3 | 24.4 | 24.4 | 12.9 |
| 2019 | 177 | 6919 | 2480 | 0.358 | 1.8 | 1440 | 538 | 34.3 | 23.8 | 24.4 | 12.9 |
| 2020 | 177 | 5035 | 1577 | 0.313 | 1.4 | 1335 | 636 | 34.6 | 24.5 | 24.5 | 12.9 |
| 2021 | 177 | 7456 | 2617 | 0.351 | 2 | 1360 | 455 | 34.4 | 24.3 | 24.5 | 12.9 |
| 2022 | 177 | 8101 | 2924 | 0.361 | 1.7 | 1421 | 377 | 35.7 | 25 | 24.7 | 12.9 |
| 2023 | 177 | 7801 | 2658 | 0.341 | 2.1 | 1532 | 245 | 34.7 | 23.8 | 24.4 | 12.9 |
| 2024 | 177 | 8353 | 2961 | 0.354 | 2 | 1489 | 222 | 35.4 | 24.4 | 24.6 | 12.9 |
| 2025 | 177 | 6502 | 2161 | 0.332 | 2 | 1397 | 435 | 34.7 | 24.6 | 24.6 | 12.9 |
| 2026 | 177 | 6550 | 2280 | 0.348 | 1.5 | 1461 | 314 | 35.9 | 25.2 | 24.7 | 13 |
| 2027 | 177 | 7469 | 2642 | 0.354 | 1.9 | 1300 | 620 | 34.9 | 24.4 | 24.6 | 12.9 |
| 2028 | 177 | 6178 | 1980 | 0.321 | 1.5 | 1710 | 582 | 33.5 | 23.3 | 24.2 | 12.9 |
| 2029 | 177 | 7459 | 2578 | 0.346 | 1.6 | 1691 | 396 | 34.9 | 23.7 | 24.1 | 12.9 |

| | | | | | | | | | | | |
|---|---|---|---|---|---|---|---|---|---|---|---|
| 2030 | 177 | 7234 | 2501 | 0.346 | 2.2 | 1488 | 277 | 35.4 | 25.3 | 24.7 | 13 |
| 1981 | 192 | 7574 | 2478 | 0.327 | 2 | 1509 | 222 | 35.2 | 23.4 | 18.7 | 12.6 |
| 1982 | 192 | 7085 | 2543 | 0.359 | 2 | 1470 | 205 | 35.9 | 23.5 | 19.5 | 12.6 |
| 1983 | 192 | 7574 | 2546 | 0.336 | 2.5 | 1390 | 285 | 34.5 | 24.2 | 17.7 | 12.6 |
| 1984 | 192 | 7878 | 2530 | 0.321 | 2.1 | 1671 | 294 | 32.9 | 21.9 | 17.5 | 12.5 |
| 1985 | 192 | 7061 | 2389 | 0.338 | 2.2 | 1671 | 350 | 33.8 | 22.9 | 17.8 | 12.5 |
| 1986 | 192 | 7966 | 2686 | 0.337 | 2.3 | 1478 | 192 | 35 | 22.6 | 19.6 | 12.6 |
| 1987 | 192 | 8485 | 3105 | 0.366 | 1.6 | 1566 | 51 | 38.2 | 25.1 | 20 | 12.6 |
| 1988 | 192 | 6676 | 2326 | 0.348 | 2.1 | 1624 | 589 | 33.6 | 23.4 | 17.3 | 12.5 |
| 1989 | 192 | 8577 | 2826 | 0.33 | 1.8 | 1491 | 101 | 36 | 23.1 | 19.9 | 12.6 |
| 1990 | 192 | 7902 | 2777 | 0.351 | 2 | 1447 | 306 | 34.3 | 23.5 | 18.1 | 12.6 |
| 1991 | 192 | 8214 | 2704 | 0.329 | 1.9 | 1425 | 227 | 35.5 | 22.6 | 19.7 | 12.6 |
| 1992 | 192 | 7388 | 2252 | 0.305 | 2.4 | 1660 | 301 | 33.7 | 21.4 | 18.8 | 12.5 |
| 1993 | 192 | 6546 | 2242 | 0.342 | 2 | 1607 | 616 | 32.4 | 22.2 | 16.7 | 12.4 |
| 1994 | 192 | 5335 | 1616 | 0.303 | 1.9 | 1587 | 622 | 32.2 | 21.8 | 16.7 | 12.4 |
| 1995 | 192 | 6483 | 2133 | 0.329 | 2 | 1363 | 406 | 34.2 | 24.1 | 17.3 | 12.6 |
| 1996 | 192 | 7777 | 2476 | 0.318 | 1.8 | 1386 | 205 | 34.7 | 22.8 | 18.8 | 12.6 |
| 1997 | 192 | 6314 | 2173 | 0.344 | 2 | 1593 | 501 | 33.2 | 22.1 | 18.2 | 12.5 |
| 1998 | 192 | 8297 | 2917 | 0.352 | 2.1 | 1474 | 272 | 35.4 | 25.2 | 18.1 | 12.6 |
| 1999 | 192 | 8301 | 2923 | 0.352 | 2.1 | 1537 | 141 | 36 | 23.4 | 19.7 | 12.6 |
| 2000 | 192 | 8455 | 2835 | 0.335 | 2.2 | 1577 | 73 | 36.1 | 22.2 | 20.4 | 12.6 |
| 2001 | 192 | 7316 | 2492 | 0.341 | 2.1 | 1498 | 319 | 35 | 22 | 19.8 | 12.5 |
| 2002 | 192 | 8420 | 2803 | 0.333 | 1.6 | 1512 | 62 | 36.4 | 24.2 | 19.5 | 12.6 |
| 2003 | 192 | 5416 | 1763 | 0.325 | 1.6 | 1641 | 401 | 33.5 | 22.2 | 17.9 | 12.5 |
| 2004 | 192 | 8006 | 2699 | 0.337 | 1.7 | 1427 | 164 | 35 | 23.5 | 18.8 | 12.6 |
| 2005 | 192 | 9025 | 2942 | 0.326 | 2.2 | 1503 | 239 | 34.7 | 23.5 | 18.4 | 12.6 |
| 2006 | 192 | 8821 | 2876 | 0.326 | 2.6 | 1473 | 169 | 34.6 | 23.5 | 18.3 | 12.6 |
| 2007 | 192 | 8398 | 2582 | 0.307 | 1.8 | 1446 | 147 | 35.2 | 23.3 | 18.9 | 12.6 |
| 2008 | 192 | 6797 | 2425 | 0.357 | 1.6 | 1410 | 366 | 34.4 | 23.9 | 18 | 12.6 |
| 2009 | 192 | 8266 | 2954 | 0.357 | 1.9 | 1411 | 321 | 35.8 | 23.6 | 19.1 | 12.5 |
| 2010 | 192 | 5732 | 1844 | 0.322 | 1.8 | 1313 | 616 | 33.6 | 24.2 | 16.8 | 12.6 |
| 2011 | 192 | 8417 | 2864 | 0.34 | 2.2 | 1413 | 297 | 34.1 | 23.2 | 18 | 12.5 |
| 2012 | 192 | 7643 | 2649 | 0.347 | 2.1 | 1412 | 394 | 34.4 | 23.9 | 17.5 | 12.6 |
| 2013 | 192 | 6887 | 2432 | 0.353 | 2.2 | 1362 | 512 | 33.9 | 25 | 16.6 | 12.6 |
| 2014 | 192 | 7654 | 2518 | 0.329 | 2 | 1475 | 197 | 35.6 | 24.5 | 18.4 | 12.6 |
| 2015 | 192 | 7097 | 2395 | 0.337 | 1.9 | 1486 | 178 | 35.2 | 23.9 | 18.4 | 12.6 |
| 2016 | 192 | 6962 | 2461 | 0.354 | 1.8 | 1471 | 247 | 34.7 | 24.5 | 17.8 | 12.6 |
| 2017 | 192 | 6337 | 2004 | 0.316 | 1.9 | 1623 | 406 | 33.1 | 21.8 | 22.7 | 12.5 |
| 2018 | 192 | 8337 | 2891 | 0.347 | 2 | 1438 | 294 | 34.7 | 22.9 | 23.2 | 12.6 |
| 2019 | 192 | 6746 | 2266 | 0.336 | 1.6 | 1660 | 475 | 33.5 | 22.1 | 22.9 | 12.5 |
| 2020 | 192 | 6457 | 2141 | 0.331 | 1.7 | 1395 | 518 | 34.1 | 22.9 | 23.2 | 12.5 |
| 2021 | 192 | 7004 | 2416 | 0.345 | 2.2 | 1689 | 441 | 34 | 22.6 | 23.2 | 12.5 |
| 2022 | 192 | 8167 | 2666 | 0.326 | 2.3 | 1451 | 366 | 34.7 | 22.6 | 23.2 | 12.5 |
| 2023 | 192 | 7673 | 2663 | 0.347 | 1.8 | 1545 | 178 | 34.6 | 23.1 | 23.2 | 12.5 |
| 2024 | 192 | 7632 | 2557 | 0.335 | 2 | 1798 | 211 | 34.4 | 22.1 | 22.9 | 12.5 |
| 2025 | 192 | 6856 | 2236 | 0.326 | 1.8 | 1456 | 395 | 34.5 | 23.3 | 23.4 | 12.6 |
| 2026 | 192 | 8767 | 2963 | 0.338 | 2.2 | 1472 | 229 | 35.3 | 24 | 23.5 | 12.6 |
| 2027 | 192 | 6542 | 2292 | 0.35 | 1.8 | 1392 | 589 | 34.4 | 23 | 23.2 | 12.6 |
| 2028 | 192 | 6426 | 2052 | 0.319 | 2.1 | 1657 | 382 | 32.6 | 20.9 | 22.6 | 12.4 |
| 2029 | 192 | 8412 | 2734 | 0.325 | 1.6 | 1711 | 348 | 34 | 21.1 | 22.7 | 12.4 |

| | | | | | | | | | | | |
|---|---|---|---|---|---|---|---|---|---|---|---|
| 2030 | 192 | 7692 | 2621 | 0.341 | 2.2 | 1533 | 242 | 35.1 | 23.8 | 23.4 | 12.6 |
| 1981 | 207 | 9120 | 2100 | 0.23 | 2.4 | 1791 | 136 | 33.8 | 20.4 | 18.1 | 12 |
| 1982 | 207 | 8302 | 2598 | 0.313 | 2.4 | 1795 | 148 | 35 | 21.3 | 18.7 | 12.1 |
| 1983 | 207 | 7765 | 2195 | 0.283 | 1.9 | 1688 | 265 | 33.1 | 20.6 | 17.3 | 12.1 |
| 1984 | 207 | 7737 | 1390 | 0.18 | 2 | 1725 | 202 | 31.9 | 19.2 | 17.1 | 11.9 |
| 1985 | 207 | 7490 | 1829 | 0.244 | 2.1 | 1781 | 274 | 33.3 | 20.4 | 17.6 | 12 |
| 1986 | 207 | 7488 | 2103 | 0.281 | 1.8 | 1782 | 157 | 34 | 19.8 | 18.8 | 12 |
| 1987 | 207 | 9022 | 2839 | 0.315 | 1.8 | 1535 | 51 | 37.3 | 23.5 | 19.5 | 12.3 |
| 1988 | 207 | 7725 | 2129 | 0.276 | 2.1 | 1660 | 505 | 33.1 | 21.3 | 17.1 | 12.1 |
| 1989 | 207 | 9571 | 2860 | 0.299 | 2 | 1830 | 82 | 34.9 | 20.6 | 19.2 | 12.1 |
| 1990 | 207 | 8747 | 2772 | 0.317 | 2.3 | 1698 | 219 | 33.5 | 20.7 | 17.9 | 12.1 |
| 1991 | 207 | 7804 | 1437 | 0.184 | 1.8 | 1775 | 167 | 34.4 | 20.2 | 18.9 | 12.1 |
| 1992 | 207 | 6876 | 1516 | 0.22 | 2.3 | 1608 | 276 | 32.8 | 19 | 18.1 | 12 |
| 1993 | 207 | 8040 | 1532 | 0.191 | 2.2 | 1700 | 251 | 32 | 20.5 | 16.6 | 12 |
| 1994 | 207 | 6188 | 1128 | 0.182 | 1.9 | 1573 | 476 | 31.4 | 19.4 | 16.5 | 11.9 |
| 1995 | 207 | 5884 | 1917 | 0.326 | 1.7 | 1649 | 403 | 33.3 | 21.3 | 17 | 12.1 |
| 1996 | 207 | 8024 | 2008 | 0.25 | 2 | 1696 | 190 | 33.7 | 20.1 | 18.2 | 12.1 |
| 1997 | 207 | 7630 | 1562 | 0.205 | 2.2 | 1699 | 385 | 31.3 | 19.4 | 17 | 11.9 |
| 1998 | 207 | 8324 | 2659 | 0.319 | 1.7 | 1375 | 335 | 34.1 | 23.6 | 17 | 12.2 |
| 1999 | 207 | 8741 | 2573 | 0.294 | 2.3 | 1548 | 104 | 35.7 | 21.2 | 19.4 | 12.2 |
| 2000 | 207 | 9266 | 2270 | 0.245 | 1.7 | 1840 | 27 | 36.2 | 19.9 | 20.3 | 12.1 |
| 2001 | 207 | 6764 | 1879 | 0.278 | 1.9 | 1731 | 307 | 34.7 | 20.3 | 19.2 | 12.1 |
| 2002 | 207 | 9166 | 2968 | 0.324 | 1.7 | 1793 | 62 | 35.1 | 21.7 | 18.7 | 12.1 |
| 2003 | 207 | 8398 | 1696 | 0.202 | 2.3 | 1747 | 169 | 33.3 | 19.9 | 17.9 | 12 |
| 2004 | 207 | 8324 | 2303 | 0.277 | 1.8 | 1746 | 164 | 33.4 | 20.2 | 18.1 | 12 |
| 2005 | 207 | 8049 | 2390 | 0.297 | 1.6 | 1781 | 237 | 33.8 | 20.7 | 18.1 | 12.1 |
| 2006 | 207 | 9279 | 2852 | 0.307 | 2 | 1761 | 117 | 33.9 | 21.4 | 17.9 | 12.1 |
| 2007 | 207 | 8199 | 1584 | 0.193 | 1.9 | 1776 | 135 | 34.4 | 20.9 | 18.4 | 12.1 |
| 2008 | 207 | 7747 | 2582 | 0.333 | 2.2 | 1737 | 230 | 33.9 | 21.7 | 17.6 | 12.1 |
| 2009 | 207 | 9620 | 2792 | 0.29 | 1.9 | 1683 | 283 | 35 | 21.4 | 18.8 | 12.1 |
| 2010 | 207 | 7300 | 2425 | 0.332 | 2.3 | 1629 | 412 | 32.8 | 21.7 | 16.5 | 12.1 |
| 2011 | 207 | 8561 | 2349 | 0.274 | 2.3 | 1679 | 237 | 33.4 | 20.8 | 17.5 | 12.1 |
| 2012 | 207 | 7943 | 2065 | 0.26 | 2.1 | 1674 | 321 | 33 | 21 | 17 | 12.1 |
| 2013 | 207 | 5436 | 1814 | 0.334 | 1.6 | 1665 | 456 | 33 | 22.9 | 16.1 | 12.1 |
| 2014 | 207 | 8649 | 2761 | 0.319 | 1.9 | 1484 | 189 | 34.6 | 22.8 | 17.7 | 12.2 |
| 2015 | 207 | 8113 | 2612 | 0.322 | 2.2 | 1797 | 86 | 34.4 | 22.1 | 17.9 | 12.1 |
| 2016 | 207 | 7438 | 2359 | 0.317 | 1.7 | 1489 | 169 | 34.5 | 22.7 | 17.5 | 12.2 |
| 2017 | 207 | 7256 | 1915 | 0.264 | 2.4 | 1605 | 358 | 32.5 | 19.2 | 21 | 12 |
| 2018 | 207 | 7552 | 1981 | 0.262 | 1.6 | 1713 | 278 | 33.1 | 20 | 21.1 | 12 |
| 2019 | 207 | 7186 | 2124 | 0.296 | 1.9 | 1721 | 398 | 32.3 | 19.3 | 20.9 | 12 |
| 2020 | 207 | 8449 | 2400 | 0.284 | 2.3 | 1697 | 271 | 33.4 | 20.6 | 21.5 | 12.1 |
| 2021 | 207 | 6340 | 1994 | 0.314 | 1.6 | 1715 | 384 | 33.2 | 20.3 | 21.5 | 12.1 |
| 2022 | 207 | 5711 | 1119 | 0.196 | 1.8 | 1681 | 301 | 33.1 | 19 | 20.6 | 11.9 |
| 2023 | 207 | 8580 | 2647 | 0.308 | 2 | 1795 | 126 | 33.5 | 20.5 | 21.5 | 12.1 |
| 2024 | 207 | 7877 | 1913 | 0.243 | 2.1 | 1792 | 125 | 34.2 | 20.5 | 21.4 | 12.1 |
| 2025 | 207 | 7467 | 2445 | 0.327 | 1.7 | 1744 | 232 | 33.8 | 21.4 | 21.6 | 12.1 |
| 2026 | 207 | 8122 | 2656 | 0.327 | 2.1 | 1486 | 163 | 34.8 | 22.1 | 21.9 | 12.2 |
| 2027 | 207 | 7874 | 2472 | 0.314 | 2.1 | 1680 | 367 | 33.5 | 20.4 | 21.3 | 12 |
| 2028 | 207 | 6365 | 1280 | 0.201 | 1.7 | 1733 | 356 | 32.5 | 19 | 21 | 12 |

## Appendix 2

Predicted and simulated yield. Predicted yield generated from Random forest algorithms and Simulated yield generated from Crop simulation model.

| Year | Sowing data | Prediction | Simulated |
|---|---|---|---|
| 1982 | 25-Jun | 2704 | 2715 |
| 1988 | 25-Jun | 2503 | 2749 |
| 1993 | 25-Jun | 2116 | 1746 |
| 1996 | 25-Jun | 2663 | 2823 |
| 1998 | 25-Jun | 2378 | 2321 |
| 1999 | 25-Jun | 2774 | 2941 |
| 2003 | 25-Jun | 2333 | 2412 |
| 2004 | 25-Jun | 2724 | 2777 |
| 2018 | 25-Jun | 2566 | 2539 |
| 2020 | 25-Jun | 2226 | 1577 |
| 2021 | 25-Jun | 2480 | 2617 |
| 2022 | 25-Jun | 2616 | 2924 |
| 2023 | 25-Jun | 2649 | 2658 |
| 2026 | 25-Jun | 2411 | 2280 |
| 2027 | 25-Jun | 2499 | 2642 |
| 1982 | 10-Jul | 2656 | 2543 |
| 1983 | 10-Jul | 2577 | 2546 |
| 1988 | 10-Jul | 2202 | 2326 |
| 2001 | 10-Jul | 2548 | 2492 |
| 2003 | 10-Jul | 2156 | 1763 |
| 2007 | 10-Jul | 2695 | 2582 |
| 2008 | 10-Jul | 2387 | 2425 |
| 2015 | 10-Jul | 2603 | 2395 |
| 2016 | 10-Jul | 2600 | 2461 |
| 2019 | 10-Jul | 2191 | 2266 |
| 2020 | 10-Jul | 2277 | 2141 |
| 2021 | 10-Jul | 2415 | 2416 |
| 2024 | 10-Jul | 2551 | 2557 |
| 2027 | 10-Jul | 2314 | 2292 |
| 2029 | 10-Jul | 2469 | 2734 |
| 1989 | 25-Jul | 2427 | 2860 |
| 1991 | 25-Jul | 2143 | 1437 |
| 1993 | 25-Jul | 2053 | 1532 |
| 1994 | 25-Jul | 1902 | 1128 |
| 1999 | 25-Jul | 2548 | 2573 |
| 2004 | 25-Jul | 2107 | 2303 |
| 2007 | 25-Jul | 2288 | 1584 |
| 2013 | 25-Jul | 2110 | 1814 |
| 2015 | 25-Jul | 2547 | 2612 |
| 2017 | 25-Jul | 1975 | 1915 |
| 2018 | 25-Jul | 1986 | 1981 |
| 2019 | 25-Jul | 2021 | 2124 |
| 2020 | 25-Jul | 2297 | 2400 |

## References


1. E.M. Abdel-Rahman, F.B. Ahmed, R. Ismail, Random forest regression and spectral band selection for estimating sugarcane leaf nitrogen concentration using EO-1 hyperion hyperspectral data. Int. J. Remote Sens. **34**(2), 712–728 (2013)
2. L. Breiman, Random forests. Mach. Learn. **45**(1), 5–32 (2001)
3. J. Chen, M. Li, W. Wang, Statistical uncertainty estimation using random forests and its application to drought forecast. Math. Probl. Eng. (2012)
4. E. Craig, F. Huettmann, Using "blackbox" algorithms such as TreeNet and random forests for data-mining and for finding meaningful patterns, relationships and outliers in complex ecological data: an overview, an example using G, in *Intelligent Data Analysis: Developing New Methodologies Through Pattern Discovery and Recovery* (IGI Global, 2009), pp. 65–84
5. G. De'ath, K.E. Fabricius, Classification and regression trees: a powerful yet simple technique for ecological data analysis. Ecology **81**(11), 3178–3192 (2000)
6. Y. Everingham, G. Inman-Bamber, J. Sexton, C. Stokes, A dual ensemble agroclimate modelling procedure to assess climate change impacts on sugarcane production in Australia. Agric. Sci. **6**(08), 870–888 (2015)
7. Y. Everingham, C. Smyth, N. Inman-Bamber, Ensemble data mining approaches to forecast regional sugarcane crop production. Agric. For. Meteorol. **149**(3–4), 689–696 (2009)
8. S. Fukuda, W. Spreer, E. Yasunaga, K. Yuge, V. Sardsud, J. Müller, Random forests modelling for the estimation of mango (*Mangifera indica* L. cv. chok anan) fruit yields under different irrigation regimes. Agric. Water Manag. **116**, 142–150 (2013)
9. J. Garca-Gutirrez, F. Martnez-lvarez, A. Troncoso, J. Riquelme, A comparison of machine learning regression techniques for LiDAR-derived estimation of forest variables. Neurocomputing **167**, 24–31 (2015)
10. J.W. Jones, G. Hoogenboom, C.H. Porter, K.J. Boote, W.D. Batchelor, L. Hunt, P.W. Wilkens, U. Singh, A.J. Gijsman, J.T. Ritchie, The DSSAT cropping system model. Eur. J. Agron. **18**(3–4), 235–265 (2003)
11. M. Lal, K. Singh, L. Rathore, G. Srinivasan, S. Saseendran, Vulnerability of rice and wheat yields in nw india to future changes in climate. Agric. For. Meteorol. **89**(2), 101–114 (1998)
12. A. Philibert, C. Loyce, D. Makowski, Prediction of $N_2O$ emission from local information with random forest. Environ. Pollut. **177**, 156–163 (2013)

# Data Analytics Using Satellite Remote Sensing in Healthcare Applications

**Kamaljit I. Lakhtaria and Sailesh S. Iyer**

**Abstract** Water Management is the greatest challenge facing mankind. Satellites remote sensing can be one of the most important sources of identifying and segregating lakes, rivers and oceans. Data Mining and Image processing are integral portions which can lead to better water management. Images through satellite of water sources like Lakes, Rivers, Oceans help identify the clusters, depth and purity. Different parameters for ensuring high quality of water include its pH balance, Acidity, Biological Oxygen demand, Hardness, Temperature etc. Satellite Images of Lakes, Oceans and Rivers can lead to conversion of pictorial data into essential parameters which can again lead to qualitative and quantitative data. Clustered Images of related patterns based on certain similarity parameters like distance, comparative clustering based on above parameters like pH Balance, Acidity, BOD, Hardness, temperature etc. This identification can lead to development of an effective model for Water and Healthcare Management using Data Mining and Analytics on images captured. These images are compared and changes recorded from time to time. Historical data can also play a vital role in predicting water sources and its capacity. Swarm Intelligence can be applied and exploited to ensure effective and quality transmission of images. The image processing and interpretation can be effectively managed as there are many images with various orientations. This study can also be used for other applications like Land Usage Statistics, Population Distribution, Farming Land Identification etc. just to name a few. Ant Colony optimization (ACO) and Particle Swarm Optimization (PSO) can be techniques of Swarm Intelligence which can be used for optimization. Data Visualization can be performed to project accurate picture of the different stages and results. Visualization of data can lead to various indicators and a dashboard made can provide all the variance of key indicators. These can also be managed by water sanitation and distribution mechanisms. Pure and drinkable or close to consumable water can be identified, purification process done through Water treatment plants

K. I. Lakhtaria (✉)
Rollwala Computer Centre, Gujarat University, Ahmedabad, India
e-mail: kamaljit.ilakhtaria@gmail.com

S. S. Iyer
Marwadi Education Foundation Group of Institutions, Rajkot, India
e-mail: drsaileshiyer@gmail.com



A. E. Hassanien et al. (eds.), *Machine Learning and Data Mining in Aerospace Technology*, Studies in Computational Intelligence 836,
https://doi.org/10.1007/978-3-030-20212-5_7

and supplied to adjoining regions which are facing water scarcity most of the time. Industrial waste polluted water can be treated through special chemical treatment plants and made worth usable for washing or general purpose.

## 1 Introduction and Historical Perspective

In earlier days, Human beings used to be in a position to identify changes in the climate and ensure that proper safeguards are taken to protect the crops, people and society at large. There were at least a few elderly people in every village who used to be expert in accurate prediction of time, weather and climatic conditions. These observations used to save the mankind from calamities like flood, drought etc.

With an enormous increase in population, industrialization, pollution, sedentary lifestyle and urbanization, the ecological balance has been disturbed. Slowly the expert elderly people have also become extinct leaving us with no option but to face the wrath of the nature. The one and only saviour for mankind is Satellite technology.

Technology has transformed and revolutionized the living standards. Human beings or rather scientists developed Space Technology. The hunger of human beings to conquer and quench for knowledge of unknown or un-chartered territories led to invention of Satellites. The world's first satellite Telstar 1 was launched in 1962.

A satellite is an object in space that orbits or circles around a bigger object. There are two kinds of satellites: natural (such as the moon orbiting the Earth) or artificial (such as the International Space Station orbiting the Earth) [1] (Figs. 1 and 2).

**Fig. 1** International space station

**Fig. 2** Artificial satellite in action

## *1.1 Working of Satellite*

Artificial Satellites are made up of materials that are able to withstand direct sun rays which may cause it to expand and contract. Also the satellite should be in a position to withstand radiation of highest level. The material used should be strong enough to face all these challenges.

The materials like Kevlar normally used to make bullet proof armour and are strong enough to face temperature changes. Another material primarily used is Aluminum as it provides weight advantage i.e. it is light in weight. This also helps ensure safety of the people traveling i.e. Astronauts.

Satellites provide accurate images or indicators which can be very helpful for various applications. Satellites can detect unwarranted motion around the border making our armed forces alert regarding unauthorized and unwanted intrusion. Satellites also help us to detect unusual activity under sea or on earth surface warning against natural disaster or man made disaster.

## *1.2 Artificial Satellites Classification*

The artificial satellites are classified based on its usage domain. The classification is as follows:

- Communication Satellites: These satellites complete the rotation in one day. Such satellites are primarily used to communicate radio or television signals to and from the earth.
- Scientific Satellites: Radio, Television, Global Positioning System (GPS), IP services, mapping of pictures and image for scientific and security applications come under the preview of Scientific satellites.

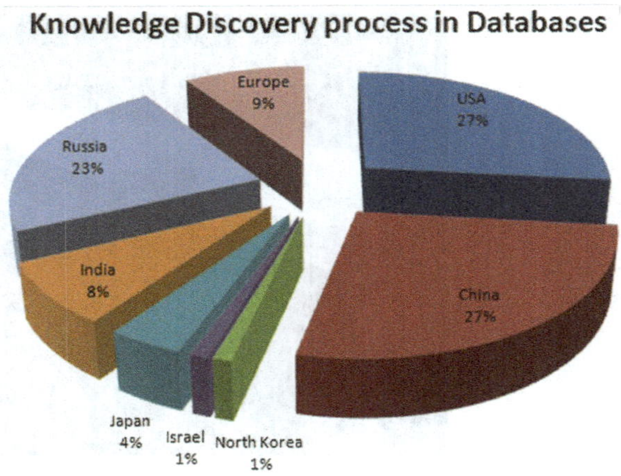


**Chart 1** Satellite orbital launch attempts country

- Weather Satellites: Some satellites are specifically used for data gathering of clouds, weather in various locations and periodic temperature alerts from various parts of the world. Weather satellites are mainly used by Weather department also known as Meteorological Department for weather analysis, prediction and issuing warnings in risky weather regions. Farmers, Fishermen etc. get timely messages so that they can plan and save their crops and life.
- Remote sensing Satellites: Such satellite can measure, observe and photograph massive land areas from space monitoring animal movement, pointing miner deposits below earth's surface, keeping a watch over the agricultural crops or the forests from being damaged and oceanic study [2].
- Special purpose Satellites: Many other satellites are used to explore sea, space exploring, astronomy, navigational purpose, search and rescue satellites etc.

Country-wise statistics of attempted orbit launches are given in Chart 1. 2014–2016 have seen an average of around 90 attempts to launch satellites in orbit.

The telemetry, tracking, and control (TT&C) subsystem of a satellite provides a connection between the satellite itself and the facilities on the ground. The purpose of the TT&C function is to ensure the satellite performs correctly, the TT&C subsystem is required for all satellites regardless of the application [3].

The three major tasks that the TT&C subsystem [3] performs to ensure the successful operation of an applications satellite:

- The monitoring of the health and status of the satellite through the collection, and processing data from the various satellite subsystems.
- The determination of the satellite's exact location through the reception, processing, and transmitting of ranging signals.
- The proper control of satellite through the reception, processing, and implementation of commands transmitted from the ground.

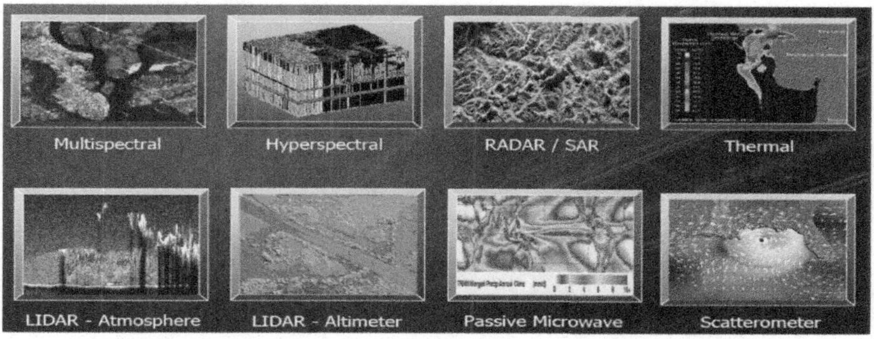


**Fig. 3** Some sample images using satellites

Telemetry is the link from satellite to base (earth) launching station. The entire dataset, images and vital information about the satellite and surroundings is provided through Telemetry.

Remote Sensing is very relevant for image capturing of various planets, oceans, rivers, maps, location and analyzing these images to develop useful content or perform Clustering, Classification or Outlier Detection.

Figure 3 denotes various sample images collected through remote sensing. The clarity, demarcation and view of these images can be useful for performing various operations for efficient utilization.

Change identification between same images taken at different duration can lead to Change Detection. Change Detection problems can be of various types:

- Binary Change Detection.
- Multiclass Change Detection.
- Changes in long time series of images.

Figure 4 shows manual thresholding and magnitude difference image depicting burned area and pixel change detection map [4]. The above views in the images provide cross-sectional view and multiple views to focus on only those images which can be helpful for analysis and decision making. Farm land development depends on assurance of water supply throughout the year and satellite images identify those areas where abundance of water can be found hence making farming convenient. These images also can give an insight into the economic condition of the farmers and neighbourhood regions. Government can roll out special schemes for those regions where water management is difficult. Crops can be stopped from being destroyed by preventing and diverting excess water to such regions where water scarcity prevails.

Water borne diseases can be prevented by anticipating the spread of these diseases due to water logging or excess collection of contaminated water coming from multiple sources into sea, ocean, river, lake etc.

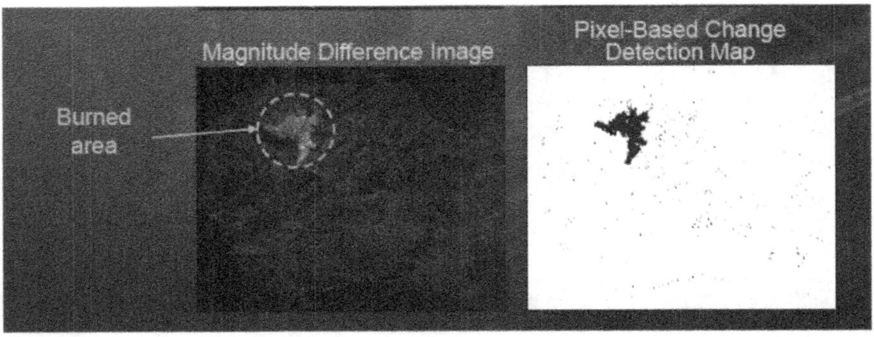


**Fig. 4** Manual thresholding

## 2 Change Detection

- Binary Change Detection:
  This change detection produces maps that are representing the areas that are changed and those that remain unchanged. This helps in detection of sudden changes or abrupt changes which can be noted and corrective steps taken (Fig. 5).
- Multiclass Change Detection: This produces change detection map which covers primarily land covered areas. They can be used for vegetation growth identification etc.
- Change in long term series:
  Change Detection in the behaviour of land between two long time series. e.g. between seasons Summer and Winter (Fig. 6).

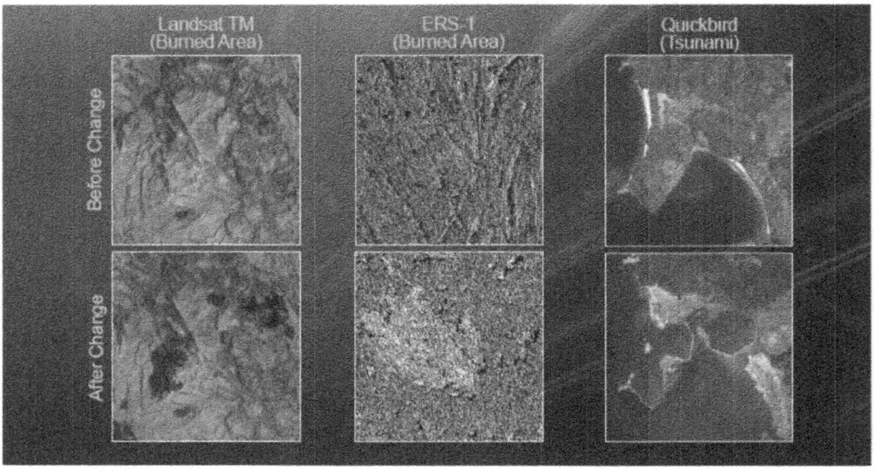


**Fig. 5** Binary change detection examples

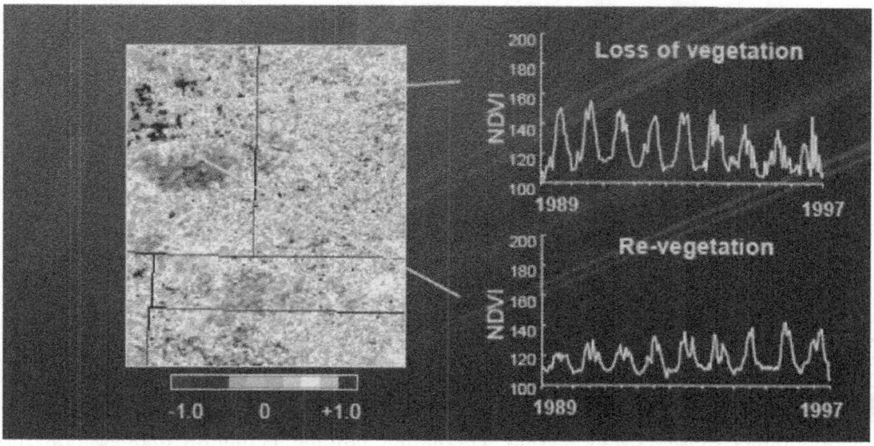


**Fig. 6** Long time series representation [5]

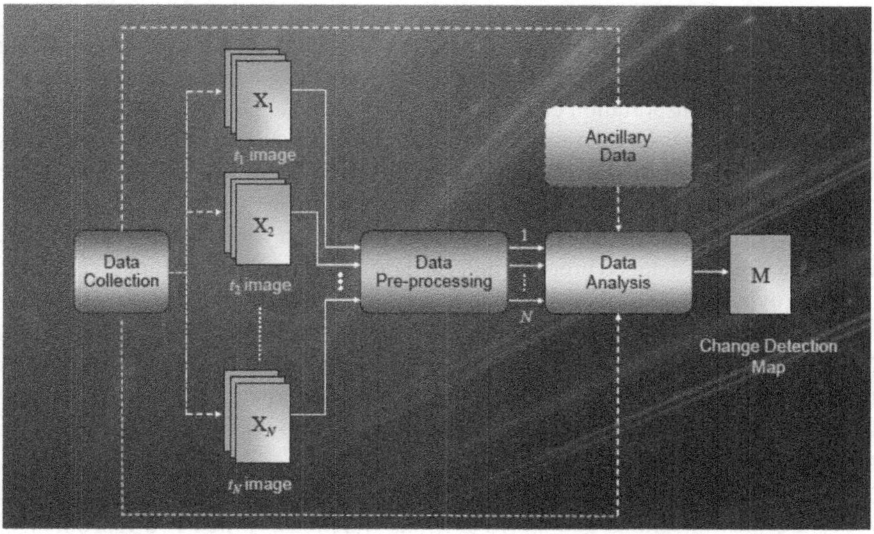


**Fig. 7** Architecture for change detection [5]

Figure 7 shows the architecture involved for detection of change.

Some common assumptions of Change Detection techniques [5] are as follows:

1. **Radiometric Issues**:
   Sensors: They should be the same for all images of same series.
   Acquisition Period: The period should be same for all applications and should be properly sampled and monitored.

Atmospheric conditions: Unnecessary portions not required should be filtered. Clouds are not required and lighting conditions need to be similar.

2. **Geometrical Issues**:
   Sensors: They should be same for all the elements of series.
   Satellite Orbit: Satellite Orbit required to be uniform either ascending or descending.
   View Angle: View angle required to be the same.

## 3 Data Pre-processing

Remote Sensing collect information about various areas from satellites. They generate abundance of data every second. The main challenge is that the data generated is in the form of digital images or videos which are consuming lot of space leading to limitations in storing for a long duration. The collected data need to be preprocessed, cleaned and made worth using.

Remote sensing datasets available from various earth orbiting satellites are being used extensively in various domains including in civil engineering, water resources, earth sciences, transportation engineering, navigation etc. Google Earth has further made access to high spatial resolution remote sensing data available to non-experts with great ease. Knowledge of Digital Image Processing of satellite data allows to process raw satellite images for various applications [6].

First the data obtained from the satellites may be in form of images or video. Images need to be mapped properly and categorized into particular zones, countries, states, province etc. The data may not be complete and may miss some vital information. It is preprocessed or cleaned to remove the noise or silence from the data. Noise refers to unnecessary data or images which may be no longer required for any decision making. Silence means some vital or important data which is missing like some image or video may not contain region information for which the image representation is given (Fig. 8).

Once the data is cleaned and preprocessed, the data from various regions and zones are then integrated and stored in Data Warehouse which is a repository of huge historical data. The data once stored in Data Warehouse cannot be updated.

The task relevant data is then segregated into Data Mart according to domain or sub-domain. These data marts are then mined for feature extraction and pattern mining leading to knowledge discovery in data. This knowledge in terms of patterns are then passed on to stakeholders so that they can utilize it to the optimum (refer to Fig. 9) [7].

**Dealing with Radiometric Issues**:
Radiometric Calibration is applied to images using two approaches:

1. Absolute Calibration:
   Digital numbers are transformed into corresponding ground reflectance values.
2. Relative Calibration:
   This approach modifies the histogram.

## 4 Data Mining and Its Techniques

Data Mining is used to extract patterns leading to knowledge from data collected from heterogeneous sources. It consists of confluence of many areas such as Statistics, DBMS, Data Warehouse, High Performance Computing, Information retrieval, Algorithms, Visualization, Pattern Recognition, Machine Learning and many more. Data Mining can be applied to many interdisciplinary domains to extract knowledge which can be useful in decision making and implementation.

Mining of images can be performed using tools like Weka, RapidMiner, R Studio, Orange and other tools like MATLAB, Scilab etc. Data Mining Techniques like Clustering, Classification and Outlier Detection are major techniques which is preferred for prediction of clusters or classification of images based on some predefined rules or to detect rivers which are totally different from other ones. In Clustering,

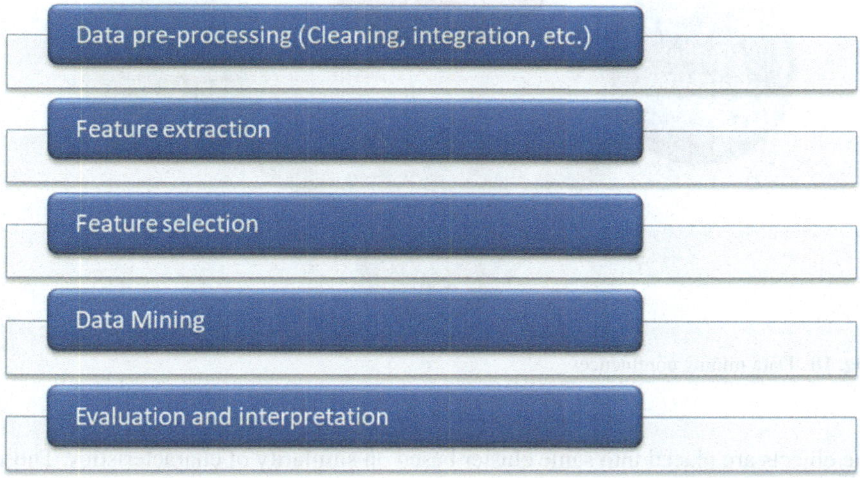


**Fig. 8** Knowledge discovery process in databases

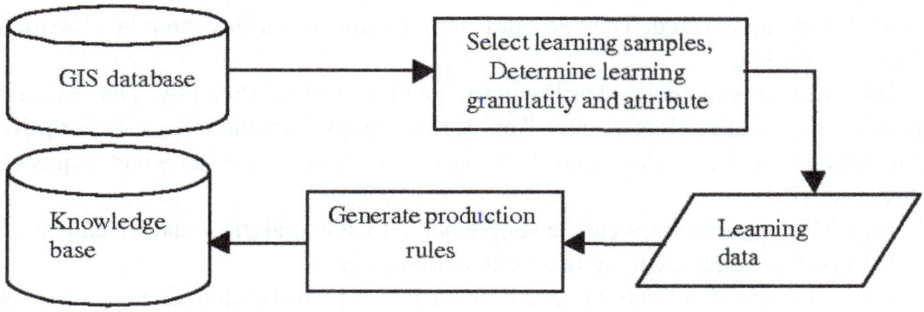


**Fig. 9** Data mining process on GIS database

**Fig. 10** Data mining confluences

the objects are placed into same cluster based on similarity of characteristics. Those objects which are not similar are forming a separate cluster. Classification technique is based on certain rules which are predefined and according to certain rules like Naïve Bayes Classifier, classification occurs. Outlier Detection refers to those elements which are isolated. They do not form into any cluster or cannot be classified (Figs. 10 and 11).

Data Mining is categorized as Predictive and Descriptive. Predictive Data Mining include Classification, Regression, Time Series Analysis, Prediction etc. Descriptive Data Mining consists of Clustering, Summarization, Association Rules and Sequence Discovery [8].

Data Mining techniques can be supplemented with Analytical and Visualization tools to perform data dredging in an efficient manner.

Clustering can be effectively used for identifying similar characteristics among rivers, lakes and oceans. All such regions having similar characteristics can be placed into same cluster and then analysis and Data Visualization can be performed. Clus-

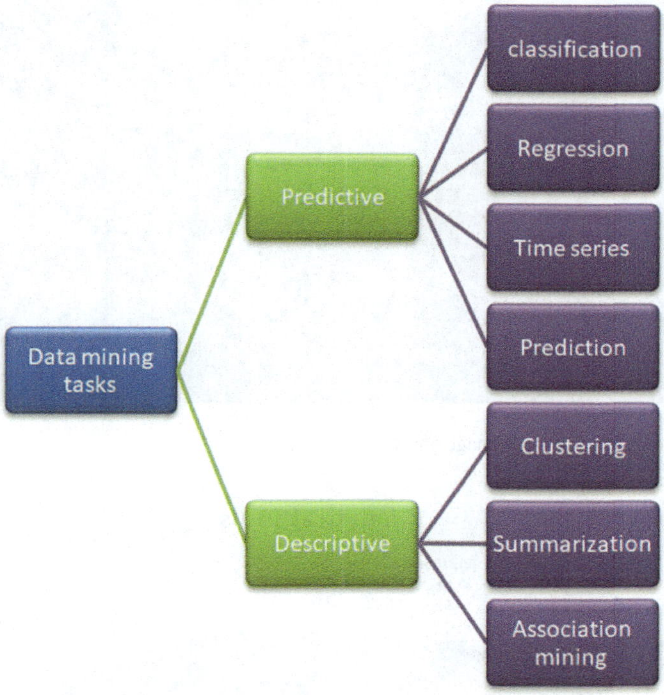


**Fig. 11** Data mining at a glance

tering mechanisms like k-Means, k-Mediods can be used to perform clustering and identify effective clusters.

Classification can also be used to classify objects based on certain rules. Naïve Bayes Classifier, Decision Tree can be mechanisms used to demonstrate the rule specific objects.

Bayesian Framework is shown in Fig. 12. There are two possible methods of Image Analysis in Bayesian Framework. They are as follows:

1. Pixel based:
   Pixel based can further be classified and Direct detection and Explicit Estimation in statistical terms using EM algorithm.
2. Context based.
   The context based analysis is based on regularization strategy and is characterized by Markov Random Fields (MRF) (Figs. 13 and 14) [9].

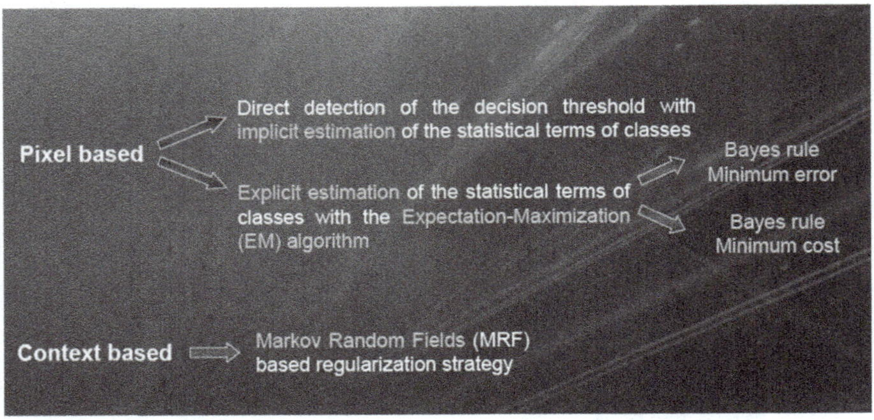


**Fig. 12** Bayesian framework image analysis

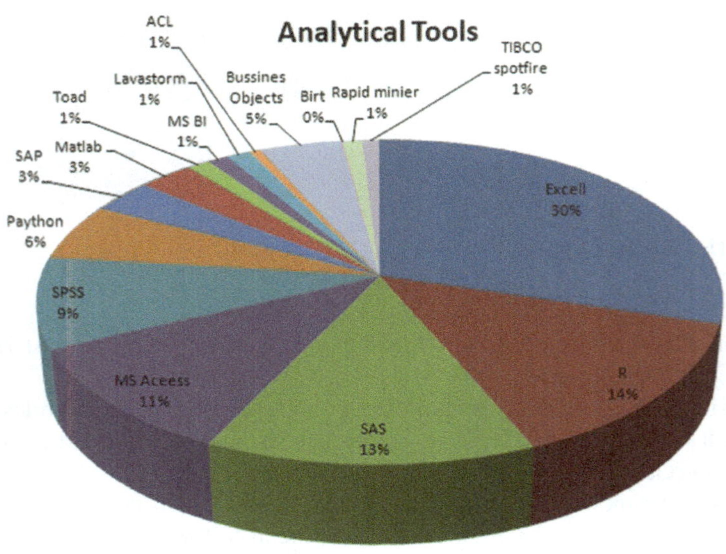


**Fig. 13** Analytical tools used as per respondent's survey

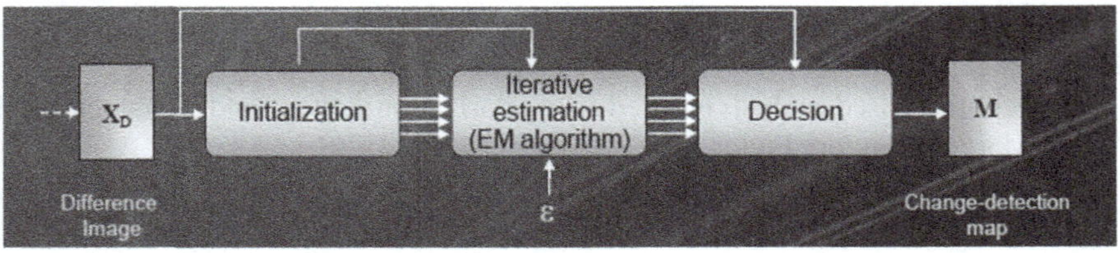


**Fig. 14** Expected maximization technique

### 4.1 *Bayesian Framework*

The solution to this problem is as follows:

1. Initialize problem and model definition.
2. Find the image difference in statistical terms using formula in an iterative manner.
3. Take a final decision (Fig. 12).

### 4.2 *Data Visualization Tools*

In addition to Analytical Tools, Data Visualization tools are used to provide graphical insight into the data in question. Some of the tools are listed and discussed below:

1. Pentaho Business Analytics.
2. Talend Open Studio.
3. JasperSoft BI.
4. Tableau.
5. Qlik.
6. Actuate.

Some of the Data Visualization tools evaluated are listed below:

1. **Tableau**:
   Tableau uses Hive to structure the queries, then tries its best to cache as much information in memory to allow the tool to be interactive. Tableau offers an interactive mechanism which allows OLAP cube analysis. Components of OLAP Cube like slicing, dicing, pivot analysis are performed with effectiveness in Tableau.
2. **Pentaho**:
   Pentaho provides a comprehensive Business Intelligence platform to analyze, integrate and present data through reports and dashboards. Pentaho supports multi-level architecture, which allows embedding analytics into any workflow application like Cloud, mobile and hybrid data models.
3. **Jaspersoft**:
   - Better Reporting to stay informed and makes better decisions.
   - Accurate Analysis to spot trends and identify issues.
   - Dashboards to view the state of your business.
   - Data Integration helps to build a data-mart or warehouse (Table 1).

**Table 1** Data visualization tools comparative analysis [10]

| Data visualization tools | Pros | Cons |
|---|---|---|
| Jaspersoft | Complete BI solutions<br>Costing very low | Less used in companies<br>Below average performance and data volumes |
| Pentaho | Ranking high among available tools<br>Cost very low | Customer feedback and support below average<br>Not easy to use |
| Tableau | Customer ranking high<br>Reusability, embedding high | High maintenance/support fees<br>High governance issues |
| Qlik | Visualization analytics high<br>Easy to use<br>Strong dashboard and big data support | Not enterprise ready<br>Risk to current customers |
| Actuate | User friendly<br>Extended big data connectivity | Non-interactive<br>Not suitable for dashboards and visualization |

**Table 2** Evaluation of data mining tools [10]

| Evaluation | R | Weka | Orange | RapidMiner |
|---|---|---|---|---|
| Association rule mining | Yes | Yes | Yes | Yes |
| K-means | Yes | Yes | Yes | Yes |
| Decision tree | Yes | Yes | Yes | Yes |
| Naïve Bayes classifier | Yes | Yes | Yes | Yes |
| Time series | Yes | Yes | No | Partial |
| Text analytics | Yes | Yes | Yes | Yes |
| Big data processing | Yes | Yes | No | No |
| Visual data workflows | No | Yes | Yes | Yes |

## *4.3 Data Mining Tools*

Table 2 gives comparative evaluation of four leading Data Mining and Statistical tools like R, Weka, RapidMiner and Orange. In the proposed model, two techniques Clustering and Classification would be used. k-Means, Decision tree, Naïve Bayes, Time Series, Visual Data Workflows would be effective mechanisms or techniques to implement.

Weka is by far the best alternative tool to implement as it is effective for Association Rule Mining, K-Means, Decision Tree, Naïve Bayes Classifier, Time Series, Text Analytics, Big Data processing and Visual Data Workflows. R Studio is also used for Mining data. Packages like rattle() can give an insight into data mining.

R is good on all other techniques but is not effective for Visual Data Workflows. Orange tool does not perform well on Time Series and Big Data processing. RapidMiner does not augur well for Big Data Processing and Time Series.

Many other tools are available like MATLAB, Lab View, KNIME which can be used for this particular model. Data Visualization Tools like Pentaho, JasperSoft, Tableau, Qlik etc.

These tools when combined with effective images can be used to provide reliable results and action based on these results can be useful to society at large.

## 5 Literature Review of Related Work in Visual Data Mining

Visual Data Mining is an exceptionally effective tool which can be used in applications where images are involved. VDM uses visual interaction to allow a human user to visually extract and explore patterns in data.

Many experiments have been conducted since year 2000 but were manual methods of Data Mining and used simple tools. Visual Data Mining was not preferred as the know how required was not available. Lucieer (2004) and Lucieer and Kraak (2004) [11] developed a visualisation tool that allowed for visual interaction with the parameters of a fuzzy classification algorithm. The study showed that visualization of a fuzzy classification algorithm in a 3D feature space plot dynamically linked to a satellite image improves a user's understanding of the sources and locations of uncertainty.

A system called Immersion Information Mining was introduced in 2013 [7]. This system uses virtual reality and is based on visual analytic approach that enables knowledge discovery from EO archives. Human Machine Interface (HMI) started in 2014 which is supported by special methods that increase the information being transmitted [12].

Spatial Data Visualization can be categorized into Geometry based, Pixel based, Icon based etc. (Table 3).

## 6 Proposed Model for Remote Sensing Using Data Mining

Geographic Information Systems can be applied in various sectors such as transport, telecommunications, public utilities, environmental design, and health services; extended to domains such as Country Planning, Geology, and Soil and Forest science, Agriculture etc.

The historical data and live data from Satellites are received by remote sensing. The data is in form of Images and Videos which require huge memory and high processing capabilities. This data needs to be converted into tabular form or text form

**Table 3** Spatial data mining steps [4]

| Criteria | Feature identification | Feature comparison | Feature interpretation |
|---|---|---|---|
| Data representation | Map display, various statistical graphic, various complicated symbol expression technique, etc. | Map overlap, map parallel, multi-dimensional color model, view framework, etc. | Automated mapping technique, information space technique, etc. |
| Data operation | Interactive map and interactive statistical graphic, focusing, sequencing, animation, data exchange, brushing, etc. | Interactive map and interactive statistical graphic, view linking, color and data assignment. | Automated mapping technique, interactive map and interactive statistical graphic, focusing, view linking, etc. |

for processing purpose. Data received from the satellite are then used for Healthcare sector and various other Government sectors to improve the living standard of mankind.

The proposed model for Remote sensing in Healthcare Sector uses Data Mining, Satellite images and Healthcare sector details as initial commencement point.

1. Collect images from heterogeneous sources obtained via satellites. The rainfall data is also collected so that clustering can be done based on frequency of rainfall.
2. Swarm Intelligence is used to get accurate and collection of images.
3. Images are cleaned to remove unwanted areas like road, mud, lakebed's etc.
4. Separate images into clusters based on similar characteristic images.
5. Store these images based on similarities in different folders.
6. Identify those images where water content is visible and then process the data by converting details into text or tabular form. This conversion can take place through online tools available.
7. Map those regions where colour of water slightly changes. Clusters of Industrial zones and areas where huge level of water contamination is also identified.
8. The lakes can be classified as Pure, Average and Below average based on their purity level and waste excretion including industrial toxins.
9. Health Department is immediately informed of change in water colour and contamination levels of water.
10. An mobile application which integrates all the above steps is developed. The advantage of this application is that the common man can know what type of water he/she is drinking.
11. This application can also use Data Analytics and Mining to predict the purity of water and spread of hazardous diseases (Fig. 15).

These satellite images can be classified based on their clarity and lighting as per standards laid down for scientific experiments.

The pixel values are obtained and scattered plot is made as follows (Fig. 16).

**Fig. 15** Satellite images of lakes [4]

**Fig. 16** Rain fall chart to study nature of lake [12]

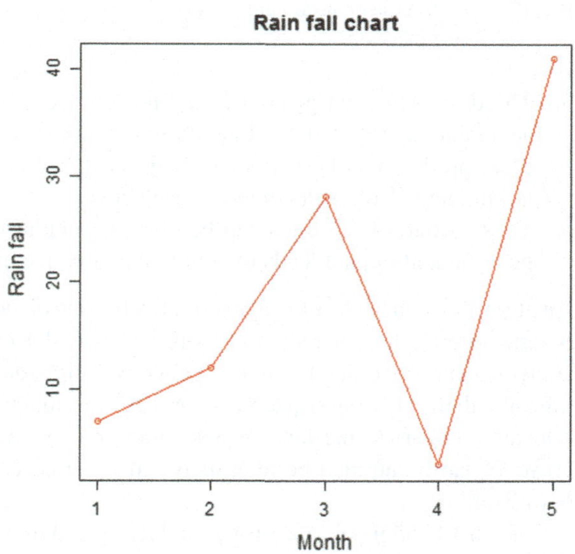


Swarm Intelligence can be effectively used to improve the content and quality of images. Swarm Intelligence is derived from the problem of Ant Colony Optimization where a collection of ants follow the same path and are able to protect themselves and optimize the path followed.

Swarms have a lot of advantages as studies have proved and some of these advantages are listed below which can be providing optimum solutions for Remote Sensing mechanism.

1. Number of Satellites flying or operating in formation giving rise to effective implementation of the proposed solution.
2. When a combination of many satellites is formed, the images obtained are multiple, having different angles of capture, better clarity and quality.

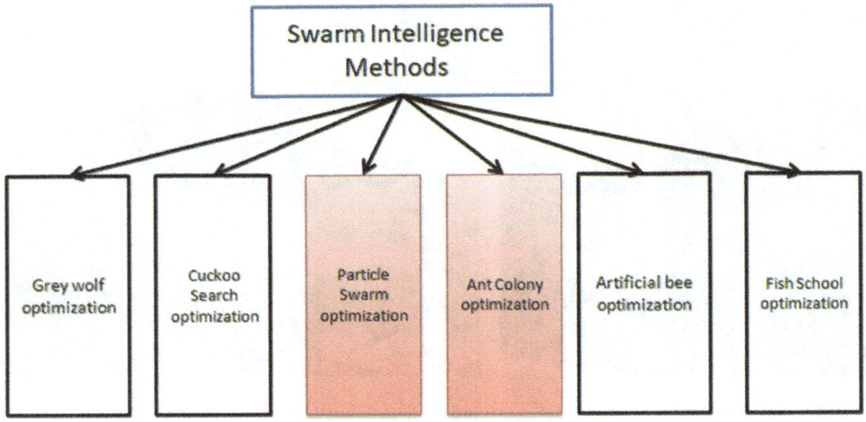


**Fig. 17** Classification of swarm intelligence

3. The design of such group of satellites ensure robustness, are autonomous and adaptable, distributed and inherently redundant.
4. Mass production of components makes it cost effective as manufacturing occurs in bulk and also results in less launch cost.
5. These group of satellites can be given particular tasks and the accuracy of the position and impact leads to better cost effective and robust solution.

An algorithm can be developed to effectively control, develop, deploy and implement Swarm Intelligence in remote sensing. The model or algorithm can be based on study of swarm intelligence and Ant Colony optimization which can capture images, transmit them at a rapid pace and process the images after classifying them into relevant categories like land, water, cities, forests etc. Sudden changes in certain crusts of earth can also be identified and reported easily with the help of Swarm optimization.

The only challenge is the proper budgeting and implementation of managing bulk components and moulding them to suit customized requirements (Fig. 17) [13].

Similarly birds move in a flock which keeps them safe and away from any hunter. This also symbolizes strength and leads to better bargain and speed. Fighter Aircraft also when on a war mission go in groups so that they can defend each other from unexpected attacks. The birds have great strength and that is their vision. They can see very sharply on both the sides. When the flock is large they cover a major portion and can view from different angles making flying in the air safe. They are also forming an interesting pattern and have impressive quality of being as a team in food searching and long distance migration.

Some of the characteristics allow birds to be very agile. They are listed below:

1. Birds are equidistant from each other and never collide with each other.
2. When they change direction immediately, they do not result in collision. The team coordination ensures that they are alert and do not collide.

**Table 4** Comparative study of ACO and PSO

| Parameter | Ant colony optimization | Particle swarm optimization |
|---|---|---|
| Problem domain | More inclined towards discrete optimization problems but also used for continuous problems | More inclined towards continuous optimization problems but now also being used for discrete problem solving |
| Representation of problem | Widely shown by weighted graph also called construction graph | Mainly shown as a set of points which are having n dimensions |
| Medium of communication | Indirect communication. Mainly ants are involved which ensure interaction through the environment | Direct interaction among particles without any change in environment |
| Where can algorithm be used? | Such problems where starting point and ending point are predefined and fixed | Mainly used where next and previous particle positions are clearly defined |
| Aim of algorithm | Searching for an optimized path in the construction graph | Finding the position of an optimized point in Cartesian coordinate system |
| Applications | Scheduling, DNA sequencing, balancing assembly lines, routing problem, travelling salesman problem | Analyze human tremor, tracking of dynamic systems, play games |

3. They avoid their enemies by changing their route or face the enemy and try to win the situation.

Ants have a habit of moving in a line following each other. They search for food and return with food particles in same order and route which they went. Since the route is tried and tested it leads to optimization as all the ants follow the same route. The food source when out of stock the ant does not leave new food pheromone trails and the volatile pheromone scent slowly evaporates. This negative feedback behaviour helps ants deal with changes in their environment [13].

There are two main approaches in Swarm Intelligence:

1. Ant Colony Optimization.
2. Particle Swarm Optimization.

A comparative study based on some common criteria can be depicted for Ant Colony Optimization vs Particle Swarm Optimization (Table 4).

The key positives of Swarm Intelligence are its Scalability, Adaptability, Robustness and Simplicity of use. Other challenges include the effectiveness of the Swarm Intelligence for time critical applications, parameter tuning, stagnation etc.

One more variant can be considered i.e. Honey Bee. They also have a unique mechanism of finding and exploiting food sources. The bee performs a particular dance to announce to other bees regarding identification of new food source. Nature has provided us with lot of guidance as far as such mechanism dealing with Swarm Intelligence are concerned.

## 7 Data Visualization and Outcomes

The proposed model outcome would incorporate the following:

1. Visualization of water resources all over the world in a systematic way.
2. Application of Clustering to identify similar lakes and atmospheric conditions.
3. Classification of lakes into various categories depending upon the level of purity, pollution, population and various other factors.
4. Distribution of surplus water in some lakes to nearby areas farms or other villages which can help the cultivation of crops.
5. Building dams in such places where drinking water can be transported to nearby town or villages.
6. Monitoring of water purity in lakes and measures can be taken to improve the quality and stop water pollution.
7. This data can be merged with Healthcare data and areas identified which are more susceptible to water borne diseases.
8. Distribution of rainfall in lakes by monitoring and taking necessary steps.
9. Government, Local Administration and public at large is aware through this application about the water level and its standards, distribution and hygiene.
10. Calamities like flood, drought can be avoided through better Water Management.
11. WASMO and such bodies can analyze and use this data to get effective results.

## 8 Conclusion

The proposed model identifies potential areas where water management is critical. Many of the lakes all over the world are in a pitiable condition. The model proposes to provide a complete satellite view of the entire lakes region wise, classify them into three categories: Pure, Acceptable and Below Average. This information would be passed on to Government Agencies of the respective regions and Local Administration for further action. Regular monitoring of health of lakes leading to better hygiene conditions would be a direct result achieved through this model.

The implementation challenges would be uninterrupted acquisition of real time images from satellite. Second challenge would be convincing local bodies to implement the findings and take appropriate steps to keep water clean. Water distribution can also be a major challenge across states and countries. Availability of healthcare statistics of different regions would also be a major challenge.

This model if implemented can be one of the best ways to keep our water clean and a great step towards hygiene and basic health care. Various other applications which can promote better water management and can help in facing conditions like drought and floods. Those areas can be identified where more water sources are available and those areas where they are scarce. Water distribution can be planned and executed in a much better way benefiting farmers, people for drinking water as water source purity and cleaning needed can be identified.

This can avoid a lot of diseases which are water borne. With this information mining, local authorities at the state level and central authorities can synchronize and prepare a work plan for better water and health management.

## References


1. https://www.space.com/24839-satellites.html
2. http://www.indiastudychannel.com/resources/149592-Artificial-Satellites-Its-Various-Types-And-Functions.aspx
3. *Presentation at Workshop on Intelligent System and Applications (ISA'17)* (Faculty of Computers and Informatics, Benha University)
4. X. Qiang, Y. Wei, Z. Hanfei, Application of visualization technology in spatial data mining, in *2010 International Conference on Computing, Control and Industrial Engineering* (2010), pp. 153–157
5. F. Bovolo, L. Bruzzone, The time variable in data fusion: a change detection perspective. IEEE Geosci Remote Sens Mag **3**(3), 8–26 (2015)
6. https://onlinecourses.nptel.ac.in/noc18_ce34/preview
7. M. Babaee, G. Rigoll, M. Datcu, Immersive interactive information mining with application to Earth observation data retrieval, in *Availability, Reliability, and Security in Information Systems and HCI*. Lecture Notes in Computer Science, vol. 8127 (Springer, Berlin, Heidelberg, 2013), pp. 376–386
8. S.S. Iyer, K.I. Kamaljit, Practical evaluation and comparative study of text steganography algorithms. Int. J. Innov. Res. Comput. Commun. Eng. **5**(3), 74–77 (2016). ISSN (Online) 2278-1021 ISSN (Print) 2319-5940
9. M. Zanetti, F. Bovolo, L. Bruzzone, Rayleigh rice mixture parameter estimation via EM algorithm images. IEEE Trans. Image Process. **24**(12), 5004–5016 (2015)
10. S.S. Iyer, K.I. Kamaljit, Practical evaluation and comparative study of big data analytical tools, in Int. J. Innov. Res. Comput. Commun. Eng. **5**(2), 57–64 (2017). ISSN (Online): 2320-9801 ISSN (Print): 2320-9798
11. A. Lucieer, M.J. Kraak, Interactive and visual fuzzy classification of remotely sensed imagery for exploration of uncertainty. Int. J. Geogr. Inf. Sci. **18**(5), 491–512 (2004)
12. D. Espinoza-Molina, M. Datcu, D. Teleaga, C. Balint, Application of visual data mining for earth observation use cases, in *ESA-EUSC-JRC 2014—9th Conference on Image Information Mining Conference: The Sentinels Era* (2014), pp. 111–114
13. https://pdfs.semanticscholar.org/116b/67cf2ad2c948533e6890a9fccc5543dded89.pdf


**Dr. Kamaljit I. Lakhtaria** is working as Associate Professor in Department of Computer Science, Gujarat University. He obtained Ph.D. in Computer Science in the area "Next Generation Networking Service Prototyping & Modeling". He holds an edge in Next Generation Network,

Web Services, Mobile Ad Hoc Networks, Network Security and Cryptography. He is author of 9 Reference Books in the area of Computer Science. He has published 3 chapters in International Editorial Volumes. He has presented many Research Papers in National and International Conferences. His papers are published in the proceedings of IEEE, Springer and Elsevier. He has 5 Ph.D. students graduate under his guidance. He is Life time member ISTE, IAENG and many Research Groups. He hold the post of Editor, Associate Editor in many International Research Journal. He is Program Committee member of many International Conferences and reviewer in IEEE WSN, Inderscience and Elsevier Journals.

**Dr. Sailesh S. Iyer** is an Associate Professor with Marwadi Education Foundation Group of Institutions—MCA Department, Rajkot. He has Ph.D. Degree in Computer Science and Research concentrated on developing and implementing an algorithm for Text Steganography. His research interests include Linguistic Steganography, Image Processing, Data Mining, Software Engineering, Project Optimization and Big Data Analytics. He is a Computer Society of India (CSI) Lifetime member and has to his credit various publications in International Journals of repute. He has also presented many Research Papers in International and National Conferences. He has served as a Judge for various events, delivered expert talks, FDP's and organized several events including AICTE sponsored National Symposium.

# Design, Implementation, and Testing of Unpacking System for Telemetry Data of Artificial Satellites: Case Study: EGYSAT1

**Sara Abdelghafar, Ahmed Salama, Mohamed Yahia Edries, Ashraf Darwish and Aboul Ella Hassanien**

**Abstract** Space industry is one of the most important industries in the modern age and used to measure the advancement of countries in the world. Egypt will launch the first satellite is designed and manufactured by Egyptian hands. In this chapter, the proposed unpacking system is developed to introduce monitoring system for the operators in the ground station through three main modules; first, unpacking module, which unpack the received packets of telemetry data from satellite to decode and display this data in readable way to the operators in the ground station. Second, limit checking module for early anomaly detection and third module is developed based on using data mining techniques for predicting the health of battery and estimate remaining useful lifetime. One of the important characteristics of this system is the flexibility of editing that makes it as a generic model compatible with any structure of cube satellite.

---

S. Abdelghafar (✉)
Computer Science Department, Faculty of Science, Al Azhar University, Cairo, Egypt
e-mail: sara.abdelghafar@yahoo.com
URL: http://www.egyptscience.net

A. Salama · M. Y. Edries
Space Division, National Authority for Remote Sensing and Space Sciences, Cairo, Egypt
URL: http://www.egyptscience.net

M. Y. Edries
URL: http://www.egyptscience.net

A. Darwish
Faculty of Science, Helwan University, Cairo, Egypt
URL: http://www.egyptscience.net

A. E. Hassanien
IT Department, Faculty of Computers & Information, Cairo University, Giza, Egypt
URL: http://www.egyptscience.net

S. Abdelghafar · A. Darwish · A. E. Hassanien
Scientific Research Group in Egypt (SRGE), Cairo, Egypt

 
A. E. Hassanien et al. (eds.), *Machine Learning and Data Mining in Aerospace Technology*, Studies in Computational Intelligence 836,
https://doi.org/10.1007/978-3-030-20212-5_8

# 1 Introduction

EgyptSat-1 (also referred to as Misrsat-1) is an international collaborative mini satellite project of NARSS (National Authority for Remote Sensing and Space Science) of Egypt and the Yuzhnoye State Design Office (YSDO), Dnepropetrosvk, Ukraine. In 2001, Yuzhnoye won the contract to design and develop the satellite, providing also technical expertise and on-the-job training to 60 Egyptian engineers and experts as well as technology transfer. EgyptSat-1 is the first remote sensing satellite of Egypt funded by the government of Egypt. The objective is to fly two instruments: a multispectral imager and an infrared imager.

Satellites are amongst today's most complex technical systems, they fulfil their mission in a very special, harsh, and challenging environment [1, 2]. So it is important to monitor the health and status of the satellite through collection, processing, and transmission of telemetry data from the various spacecraft subsystems to the ground station to ensure the satellite performs correctly, which is the major task of Telemetry, Tracking and Control (TT&C) subsystem of satellite that provides a connection between the satellite itself and the facilities on the ground. Telemetry is the collection of measurements and onboard instrument readings required to deduce the health and status of all of the satellite subsystems in the spacecraft bus and the payload. So the ground station needs a system to decode and display this data for the user in a friendly user interface program to get the performance and status of all subsystems that will help to take appropriate decision and predict any failure [3, 4] (Fig. 1).

The unpacking operation is the reverse of packing. The packing and unpacking process can be complicated. The proposed unpack system in this chapter of the telemetry packets sent by the satellite in each session and illustrate the data sent in

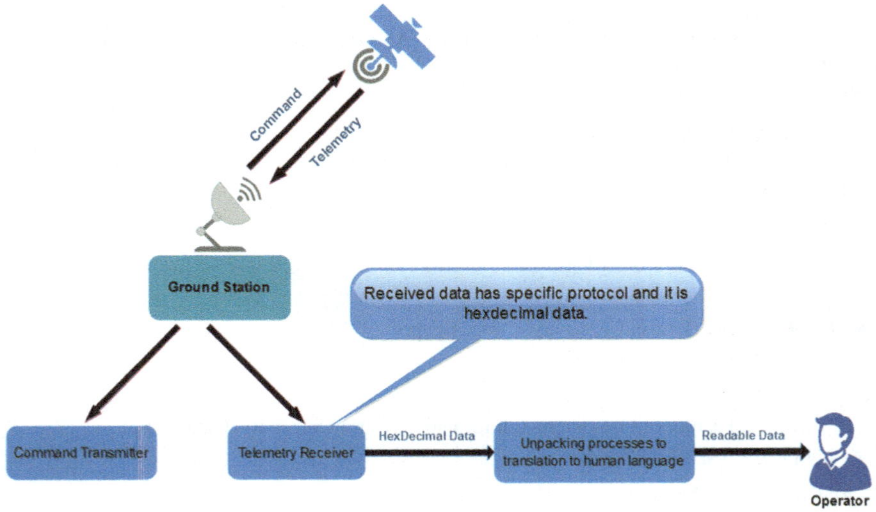


**Fig. 1** Satellite and ground control segment

readable and understandable way to the operator. In addition to this there will be warning if any sensor had a value out of its range (Minimum and Maximum values) and it will give an estimation to the battery life remaining and the duration till it may be totally damaged using data mining algorithms.

We started studying Egyptian satellite EGYSAT1 as the first test case which was jointly built by Egypt's National Authority for Remote Sensing and Space Sciences together with the Yuzhnoye Design Bureau in Ukraine, the work has been tested specifically on the power subsystem. Then we had to make a more generic telemetry unpacking program that can be compatible with any structure of cube satellites. We started changing in the core of the program (Database schema, software, GUI (Graphical User Interface)…etc.). The system features isolation, safety, data security and user-friendly GUI. The isolation comes from the OOP (Object Oriented Programming) principle where every function is done by a separate class and that ensures stability and ease of modifications. The safety comes from multi-layer architecture that secures the stored data. Data is stored in the database and can be retrieved in any time to do the needed operations on it. Finally the GUI gives the operator the ability to make an online session with the satellite or to unpack a previous stored session and the graphs also helps in visualizing the data for better understanding. In addition to this we added the data mining module to the system that learns the unpacked data and then uses it to predict the future packets which leads to the estimation of the battery state of charge and the battery life remaining. A system was created to unpack the packets of the EGYSAT1 that will be discussed later. Then a generic model had to be created that can work with any cube satellite.

This chapter is organized as follows. Section 2 presents the details of the proposed unpacking system. In this section, system characteristics, architecture, and design are presented. Section 3 presents case study of EGYSAT1 as an application for unpacking telemetry data. Section 4 concludes this chapter.

## 2 The Unpacking System

The framework of implemented system is designed to introduce three main modules:

- Unpacking module, which unpack the received packets of telemetry data from satellite to decode and display this data in readable way to the operators in the ground station.
- Limit checking module, which check sensor values are within pre-determined ranges which are specified by upper and lower limits and issue a warning if any of them is violated, that is considered one of the basic methods for early anomaly detection process.
- Mining module, which is to acquire the system behaviour models necessary for anomaly or fault detection and prediction through storing a vast amount of received data, and then processing by the machine learning techniques.

## 2.1 Unpacking Module

Applying the OOP (Object Oriented Programming) principle was useful in the project as it makes every functionality done by a single class which helps in isolation and modifications can be easily done in a single class. The main concept of the program is that it's depending on the database which makes it generic that enables the user to change in the input of the program by changing in the database not in the code to make it easier to change parameters or even the format of the packet received which means that the satellite has been changed. Figure 2 shows the main classes of the systems and its functions.

**Class Conversions**:

1. **Initialization**: it's a function that initialize the hash map structure to contain the values of the hexadecimal numbers with the corresponding binary values.
2. **ConvertHexToBinary**: it's a function to converts hexadecimal values to binary.
3. **ConvertIntToBinary**: it's a function that convert Integer numbers to binary numbers.
4. **SelectSpecificBits**: it's a function that select a specific number of bits form a bigger or equal size byte and that specific number given as a parameter.
5. **ConvertTwoSComplement**: it's a function responsible for getting the two's complement of the desired binary value.

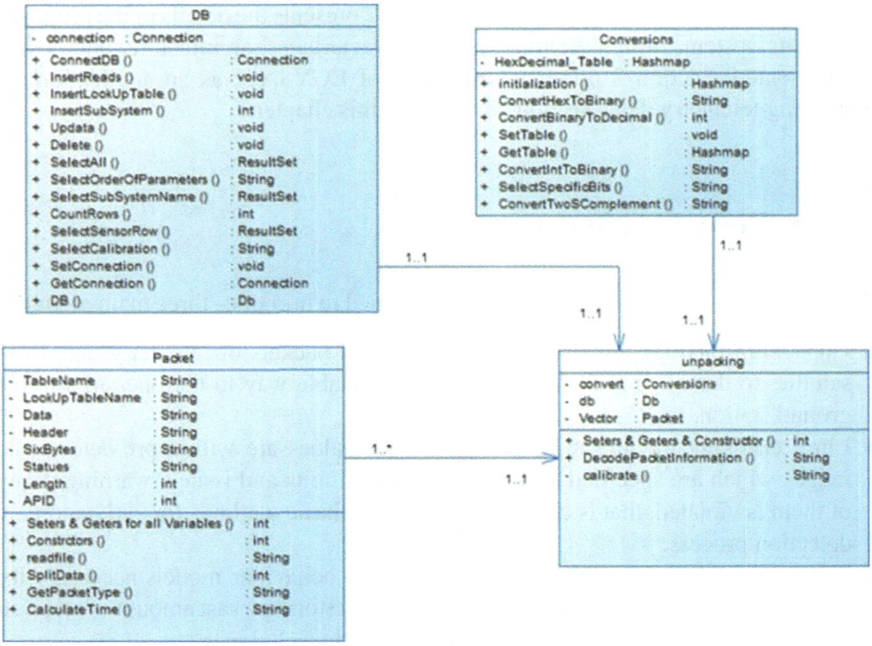


**Fig. 2** Generic module class diagram

**Class Database**:

1. ConnectDB: it's a function that establish the connection with the database.
2. InsertLookUpTable: it's a function that insert the values in the selected lookup table.
3. InsertSubSystem: it's a function the inserts a new subsystem.
4. Update: it's a function that update the value of an attribute of a table.
5. Delete: it's a function that delete a row or more from a specific table and on specific condition.
6. Select: it's a function that selects all the values in a certain table.

**Class Packet**:

1. ReadFile: it's a function that reads the frame file and loads it into the memory.
2. SplitData: it's a function that gets the APID and data length from the packet.
3. CalculateTime: it's a function that calculates the time by adding the amount of seconds found in the time packet and the preset time.

**Class Unpacking**:

1. DecodePacketInformation: it's a function that combines all the functions above and it gets the reads of each sensor in binary and converts it to decimal.
2. Calibrate: it's a function that that take the read of each sensor in decimal and gets it calibrated by its calibration factor which has many cases which can be no change, choices and equation.

### 2.1.1 System Characteristics

All of the previous functions are delivered with many distinctive characteristics, the most important of these:

- Efficiency and flexibility of editing, where the editing could be forced by the lookup tables of database without any needs for changing in the programming code that makes it as a generic model compatible with any structure of cube satellite,
- Isolation and that is done by the OOP principle which keeps every function done by a single class with no interfering between classes which ensures accuracy. It offers also the ease of modification in the software as modification can be done easily with interfering with other classes.
- Data security done by saving the unpacked packets in the database with the suitable backups. The database features the fast retrieval of the data and update or deletion of the data.

### 2.1.2 System Architecture

The System architecture is a three tier architecture consisting of three tiers as shown in Fig. 3. These tiers are presentation tier, logic tier and data tier.

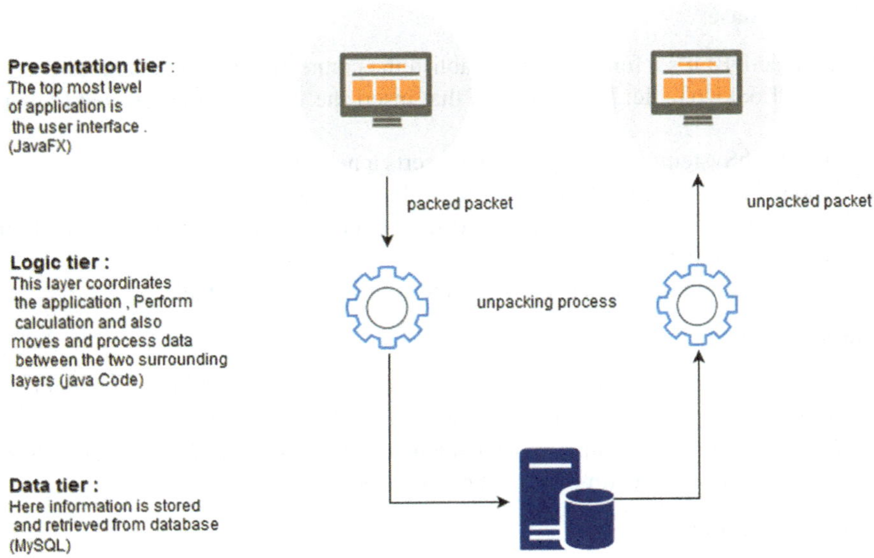


**Fig. 3** System architecture diagram

1. The data tier is the data stored in the database.
2. The logic tier is considered as the processing unit of the program responsible for the calculations and manipulating the data between the other 2 tiers. It's also the layer containing all the back end software controlling the unpacking process and the mining process as well.
3. The presentation tier is mainly the GUI which is always in contact with the operator of the program.

One of the main advantages of this architecture is the security maintained by these layers as anyone can't access the data without going through the three tiers with their security. It's also useful in the isolation property and the ease of modifications in the future.

### 2.1.3 System Design

A. Development methodology

The main idea about the proposed schema is to be generic and adaptable with any satellite. That needs the program to be flexible with the number of subsystems, different types of calibration, number of sensors in each subsystem and so on. Below is a brief description of major tables:

1. Table system is the table containing each subsystem name and description.
2. Table Packet receive is the table containing each packet ID joining it with the session ID.

3. Table Standard is the table containing the standard of the satellite which is the APID-which is the identification of which subsystem does the current packet belong to- starts from. Also the bit Data part starts from and same to the rime part.
4. Table sensors is the table containing all the used sensors with all the information about each one including the description, minimum value, maximum value, unit and so on.
5. Table storage is the table containing all the unpacked packets of all subsystems.
6. All the tables of types are the different types of calibration as example equations, on and off and limits.

### 2.1.4 Interface Architecture

User interface help operator to watch unpacking process, each sensor with his value for every sensor, making charts depend on the values stored for every sensors and also make red alarm if value of a sensor is out of his limits. Figure 4 shows the class diagram having UI classes as a view layer and control layer.

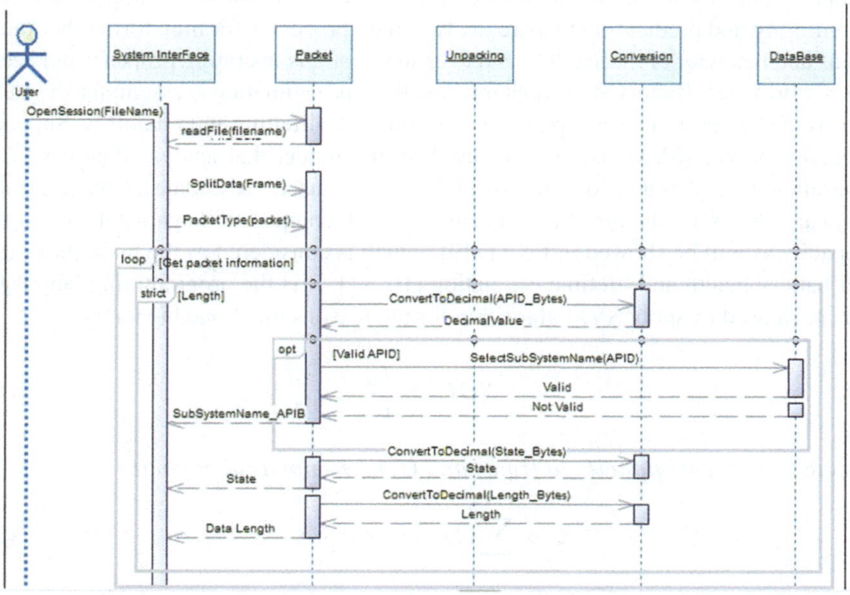


**Fig. 4** Class diagram with UI components

## 2.2 Limit Checking Module

Limit checking is the most fundamental and the most widely used anomaly detection technique for satellite systems. The reason for this method popularity is that it is easy for human operators to implement the system, apply it to the spacecraft, and understand the detection result. It constantly monitors some important time series in the telemetry data and checks whether the value is within the pre-defined upper and lower limits of various sensor values such as bus currency, voltage, angular velocity, temperature, and so on [5, 6]. This was achieved by checking on the minimum value and the maximum value of each sensor during the display of the unpacked packets. The green color is applied to the value if it is in the normal range. The red color is applied to the value if it is outside the normal range. This gives the operator an alarm when the sensor reads go out of range indicating failure.

## 2.3 Mining Module

Using data mining for telemetry data is essential to ensure that a satellite is operating properly and has no anomalies that could threaten its mission. Mining introduces monitoring and predication that are the two main processes for monitoring the functions and behavior of the satellite and to ensure that it is operating properly and keep its performance. Battery state of charge and lifetime estimation is essential to a satellite as the battery is a critical part and determines the lifetime and reliability. Support Vector machine (SVM) is a supervised learning model that analyze data used for classification and regression analysis [7, 8]. SVM is used to estimate a battery's state of charge based on the predicate capacity, through comparing it against the nominal capacity as will be showed in Eq. (1) [9], which is considered as the main indicator for battery health and lifetime estimation [10, 11]. R is the programming language has been used to apply SVM algorithm for the reasons mentioned recently.

$$SOC_t = \frac{C_t}{C_n} \tag{1}$$

where $C_t\ current\ capacity\ at\ time\ t\ and\ C_n\ is\ the\ nominal\ capacity.$

$$C = \sum (I_{disch} - I_{ch}) \cdot \Delta t \tag{2}$$

where $I_{disch}$, $I_{ch}$ the current in discharge and charge modes respectively. Equation (2) is used to calculate the capacity of the battery where the capacity can't be measured directly for in-orbit satellite, but can be calculated using the current as show in the equation [12, 13].

## 3 Case Study: EGYSAT1

EGYSAT1 consists of two main components which are the payload and satellite bus. The payload is the main mission of the satellite. The satellite bus consists of many systems which are power, communication, Attitude Determination and Control System (ADCS), telemetry, tracking and command (TT&C), thermal control, structure, onboard computer (OBC) and propulsion subsystem as shown in Fig. 5.

(1) The power system is responsible for the management of the power generated by the solar cells and stored in the batteries in the satellite to certain levels to maintain the availability of power when the satellite needs it.
(2) The communication system is responsible for the communication with the ground station on earth.
(3) ADCS is one of the most important systems in the satellite as it responsible for many tasks:
    (a) In the first launch it stabilizes the satellite by damping the angular velocity and initializing the construction of satellite attitude.
    (b) Changing the attitude and orientation needed to capture image.
    (c) Keeps the orientation to stay in touch with the command with low accuracy and low power consumption.
    (d) The execution of the ground station order of the desired attitude and orientation.

The determination of the position of the satellite is done in the ADCS system using Magnetometer, Sun Sensor and Star Sensor. In addition to this the changing in the attitude or the orientation of the satellite is done in the ADCS by the Magneto Torque (MT) and The Reaction Wheel. There several operational modes of this system which are IAA (DE tumble mode) (Damping mode), SB (Stand By mode), PTM (Imaging mode), HAAC and EM (Emergency mode).

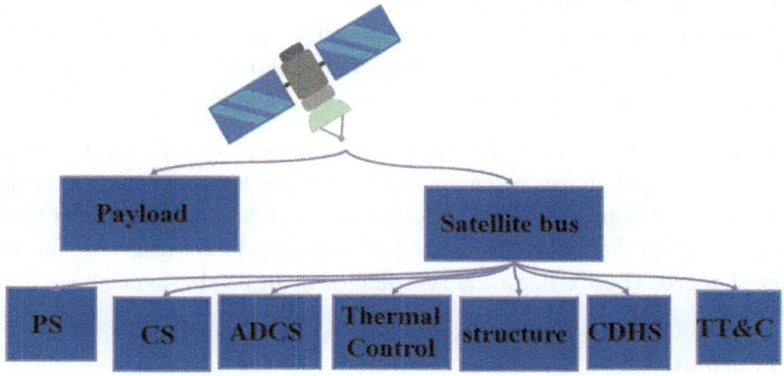


**Fig. 5** Satellite structure

(4) The Telemetry system is responsible for combining all the reads of all the system into the form of the packet sent from the satellite to the ground station.
(5) The Structure system is responsible for measuring the health of the components of the satellite.
(6) The On Board Computer System (OBC) is the brain of the satellite which manages all the systems above and it controls the satellite by the orders sent from the ground station.
(7) The propulsion system is responsible for the control of the satellite by the external thrust generated for the satellite movement to change speed or to maintain certain condition.

## *3.1 EGYSAT1 Unpacking Module Design*

The Applied system consists of three main components; database, user interface and backend software [14, 15]. Firstly the database is developed using MySql consisting of table for each subsystem. There is a table to store the raw data (unprocessed) of each subsystem and the lookup tables (packet format) of each subsystem.

Secondly, the user interface is an easy user interface which enables the user to access all the live feeding of the unpacked data of each subsystem and showing him the statistics of each subsystem represented in graphs and charts. Finally, the background is developed using java consisting of several classes. The class diagram of EGYSAT1 is shown in Fig. 6 where

(1) **Database Class**: A class to manage the database. It connects to the database and retrieves data from certain table (Lookup tables) and add the unpacked data to the storing tables.
(2) **Packet Class**: A class that splits the packet received to separate values each one represent a unique parameter.
(3) **Conversions Class**: A class that is responsible for the conversions between the decimal, hexadecimal and binary formats.
(4) **Unpacking Class**: A class that is responsible for the calculation of values of the reads of the sensors by using the values returned from the lookup table of the desired system and its calibration factor.

Applying the OOP principle, the main concept of the program is that it's depending on the database which makes it generic that enables the user to change in the input of the program by changing in the database not in the code to make it easier to change parameters or even the format of the packet received which means that the satellite has been changed. These criteria were considered in the database schema as indicated in Fig. 7.

**Packet Format**:

The EGYSAT1 telemetry packet format is described in Fig. 8.

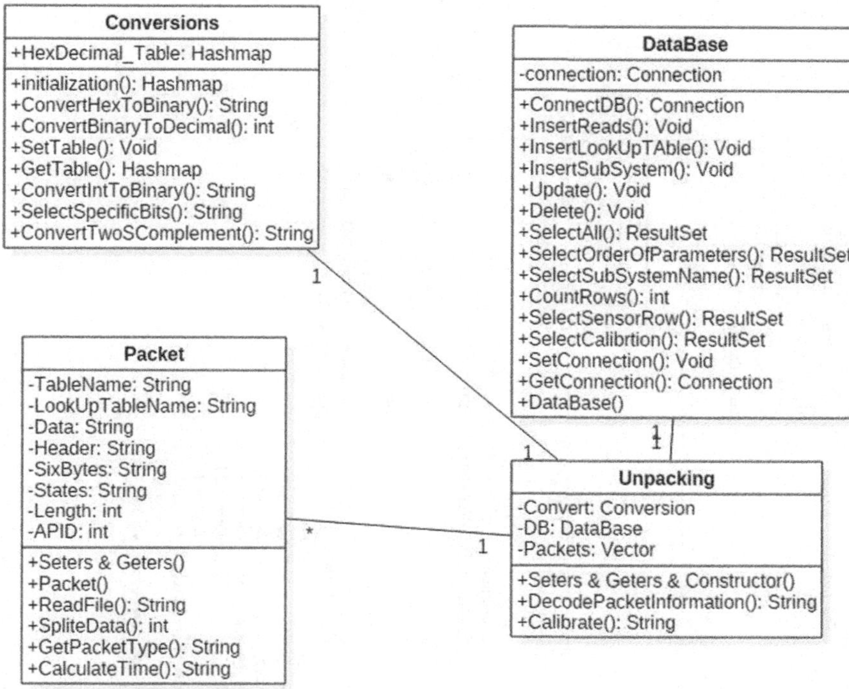


**Fig. 6** EGYSAT1 class diagram

### 3.2 *Test Data*

We used in the test data a three months telemetry data of power subsystem, which has been collected from 69 sensors as shown in Table 1. These 3 months of telemetry data made about 106,000 unpacked packet stored in the database. Figure 9 shows screenshot of unpacked packet of EGYSAT1.

## 4 Conclusion

In this chapter, unpacking system is developed to be used to monitor satellite subsystems in the ground station through three main modules; unpacking, limit checking and mining. The telemetry packets of EGYSAT1 satellite is used as a case study, the first phase of the applied system is the unpacking module that is created to unpack the received packets of telemetry data from satellite to decode and display this data in readable way to the operators in the ground station. In the second phase, limit checking and mining modules are developed for early anomaly detection and for predicting the health of battery and estimate remaining useful lifetime. From the

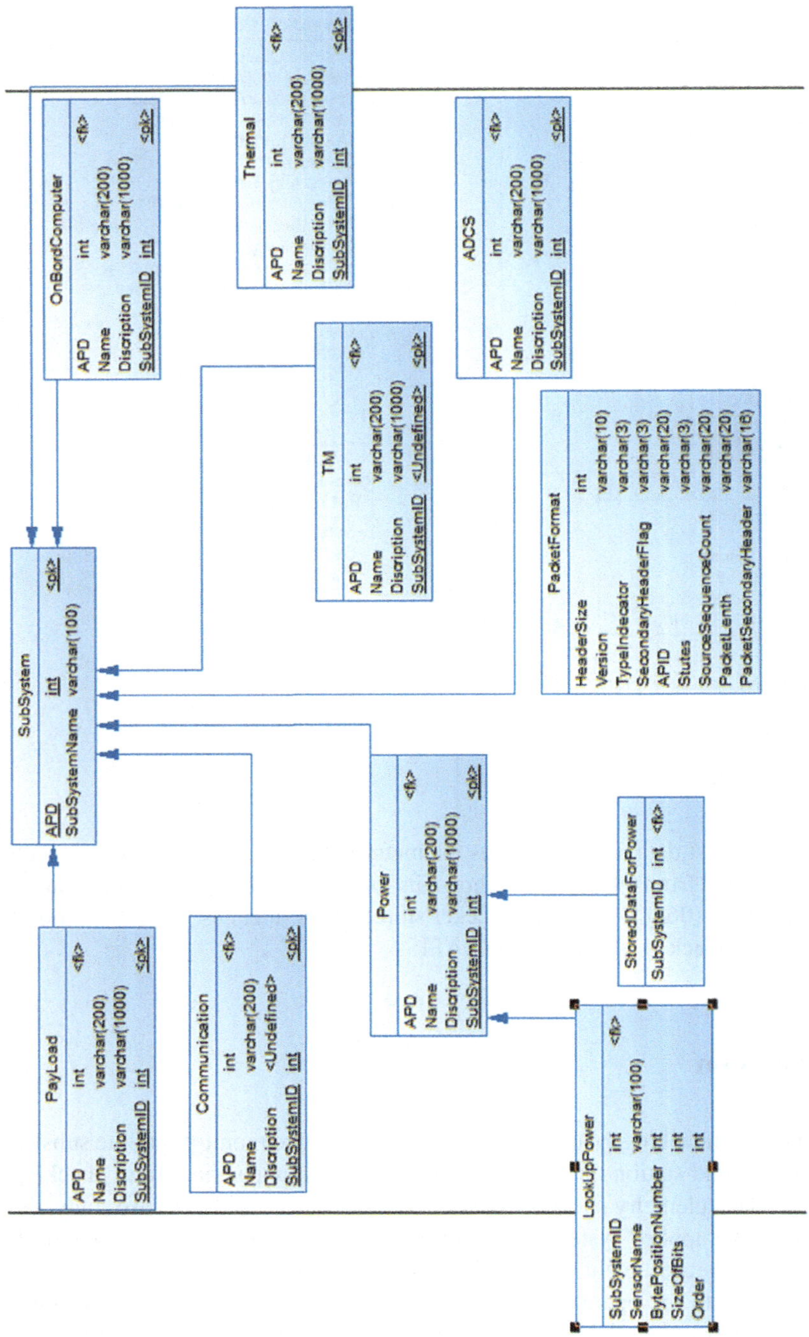


**Fig. 7** EGYSAT1 ERD

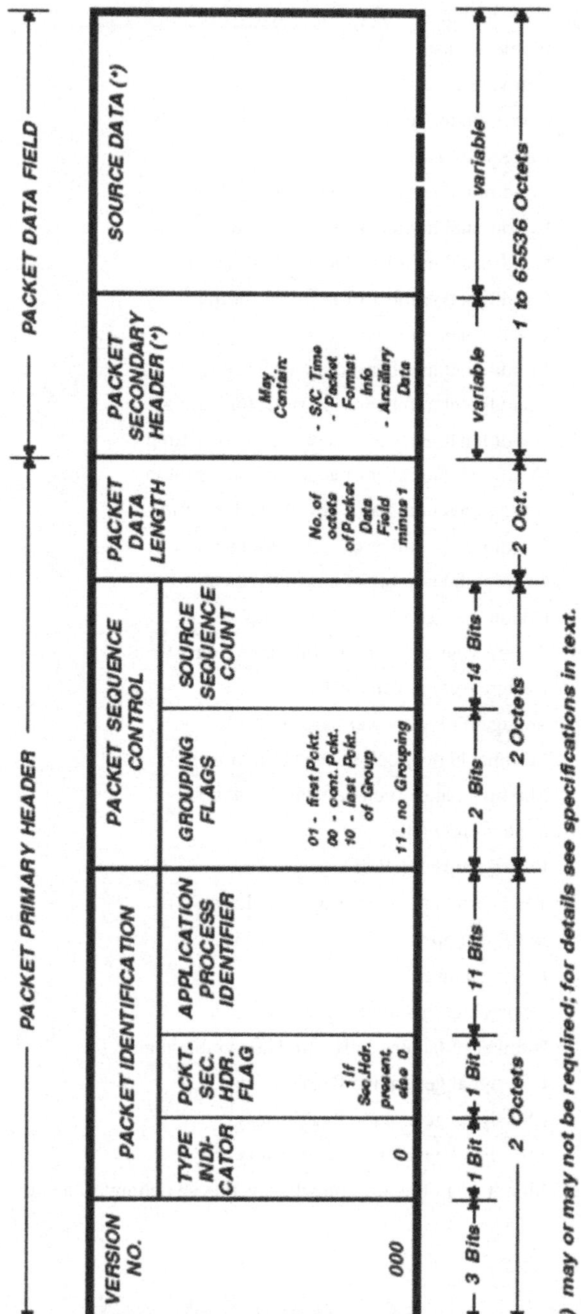


**Fig. 8** EGYSAT1 telemetry packet format

**Table 1** Sensors of the power subsystem description

| Sensor name | Description |
|---|---|
| NOM KADR | Frame number |
| VERPR | Software version |
| REJIM RAB | Operation mode |
| RHh | Battery discharge |
| USL N Ufd | Conditional number of active Uf setting |
| USL N Ufn | Conditional number of new Uf setting |
| USL N Ub1d | Conditional number of active Ub1 setting |
| USL N Ub1n | Conditional number of new Ubl setting |
| USL N Ub2d | Conditional number of active Ub2 setting |
| USL N Ub2n | Conditional number of new Ub2 setting |
| USL N atd | Conditional number of active factor setting |
| USL N atn | Conditional number of new AT factor setting |
| USL N aid | Conditional number of active AI factor setting |
| USL N ain | Conditional number of new AI factor setting |
| USL N Kpd | Conditional number of active Kp factor |
| USL N Kpn | Conditional number of active Kp factor |
| USL N Kpz | Conditional number of Kpz factor |
| RUogr | Design voltage of charge limiting (Uogr) V |
| RSUco | Design average voltage V |
| NN1 | Voltage on PSS power buses (UN) V |
| Uamin | Minimal battery cell voltage (Ua min) V |
| Uamax | Maximal battery cell voltage (Ua max) V |
| TN1 | Load current (IL) |
| TBH | Battery current (IBAT) |
| TBS1 | Total current of solar array complex (ISA) |
| T1BH | Battery temperature |
| T2BH | Battery temperature |
| TRBH | Design battery temperature |
| KCA | Number of battery cells with Ua no more than 1.1 V |
| DC1i | ES1 signal generation flag (history) |
| DC2i | ES2 signal generation flag (history) |
| DC3i | ES3 signal generation flag (history) |
| KUOVN | Monitoring of control of cells leveling switching off and on |

(continued)

**Table 1** (continued)

| Sensor name | Description |
|---|---|
| NABi | Cell leveling on flag (history) |
| NAOi | Cell leveling off flag (history) |
| KUON RN | Monitoring of control on "Load Off" (ON) and "Load On" (RN) signals |
| Oni | Load Off (ON) flag history) |
| RNi | Load On (RN) flag history) |
| OZC | Stepped charge limiting (OZC) flag |
| BON | Load Off (ON) blocking |
| BRSD | Stepped charge mode blocking |
| BSACHrk | Ampere-hour counter blocking by real-time command |
| BSACHav | Ampere-hour counter automatic blocking |
| BR OBM OZU | Refuse of data exchange with external RAM |
| TAIM1 | Operation on Int4 (Redundancy switch control unit timer 1) |
| TAIM2 | Operation on Int45 (Redundancy switch control unit timer 2) |
| TAIMvn | Operation on internal timer |
| PNBi | Battery maximal voltage (PNB) (history) |
| SNBi | Battery average voltage (SNB) (history) |
| FNBi | Battery fixed voltage (SNB) |
| MNBi | Battery minimal voltage (MNB) (history) |
| MNA1i | Battery cell minimal voltage (MNA1) (history) |
| MNA2i | Battery cell minimal voltage (MNA2) (history) |
| P MIKRZIK | Micro cycling flag |
| Uet1 | Standard source voltage (Ust1) V |
| Uet2 | Standard source voltage (Ust2) V |
| Uuu | Control unit power supply voltage V |
| TSTpu | Flash-memory test (control program) |
| TSTps | Flash-memory test (communication program) |
| PREJ | Switch to current mode condition |
| U P MIKRZIK | Switch condition at micro cycling |
| KEN | Control of load power consumption |
| N PODKONT | Controller sub-channel number |
| RHs | Battery discharge |
| USLN Con d | Conditional predetermined level number is active |
| USLN Con n | Conditional predetermined level number is new |
| USLN dUd | Conditional predetermined level number d Ud is active |
| USLN dUn | Conditional predetermined level number dUn is new |
| SKLSHREG | Control of shunt regulator switches status |

| | A | B | C | D | E | F | G | H | I | J | K | L | M | N | O |
|---|---|---|---|---|---|---|---|---|---|---|---|---|---|---|---|
| 1 | SUBSYSTEMID | Session | NOM_KADR | VERPR | REJIM_RAB | RHH | USL_N_UFD | USL_N_UFN | USL_N_UB1D | USL N UB1N | USL_N_UB2C | USL_N_UB2N | USL_N_ATD | USL_N_ATN | USL_N_AID |
| 2 | 1 | 1442 | 46 | 7 | WLM | 1.9648437500000002 A*h | 33.0 V | 33.0 V | 32.5 V | 32.5 V | 28.5 V | 28.5 V | -0.05 V/C | -0.05 V/C | 0.15000000000000002 |
| 3 | 1 | 1442 | 47 | 7 | WLM | 1.9640625000000003 A*h | 33.0 V | 33.0 V | 32.5 V | 32.5 V | 28.5 V | 28.5 V | -0.05 V/C | -0.05 V/C | 0.15000000000000002 |
| 4 | 1 | 1442 | 48 | 7 | WLM | 1.9631250000000002 A*h | 33.0 V | 33.0 V | 32.5 V | 32.5 V | 28.5 V | 28.5 V | -0.05 V/C | -0.05 V/C | 0.15000000000000002 |
| 5 | 1 | 1442 | 50 | 7 | WLM | 1.9612500000000002 A*h | 33.0 V | 33.0 V | 32.5 V | 32.5 V | 28.5 V | 28.5 V | -0.05 V/C | -0.05 V/C | 0.15000000000000002 |
| 6 | 1 | 1442 | 51 | 7 | WLM | 1.9603125000000001 A*h | 33.0 V | 33.0 V | 32.5 V | 32.5 V | 28.5 V | 28.5 V | -0.05 V/C | -0.05 V/C | 0.15000000000000002 |
| 7 | 1 | 1442 | 52 | 7 | WLM | 1.9595312500000002 A*h | 33.0 V | 33.0 V | 32.5 V | 32.5 V | 28.5 V | 28.5 V | -0.05 V/C | -0.05 V/C | 0.15000000000000002 |
| 8 | 1 | 1442 | 53 | 7 | WLM | 1.9585937500000001 A*h | 33.0 V | 33.0 V | 32.5 V | 32.5 V | 28.5 V | 28.5 V | -0.05 V/C | -0.05 V/C | 0.15000000000000002 |
| 9 | 1 | 1442 | 54 | 7 | WLM | 1.9575000000000002 A*h | 33.0 V | 33.0 V | 32.5 V | 32.5 V | 28.5 V | 28.5 V | -0.05 V/C | -0.05 V/C | 0.15000000000000002 |
| 10 | 1 | 1442 | 55 | 7 | WLM | 1.95671875 A*h | 33.0 V | 33.0 V | 32.5 V | 32.5 V | 28.5 V | 28.5 V | -0.05 V/C | -0.05 V/C | 0.15000000000000002 |
| 11 | 1 | 1442 | 56 | 7 | WLM | 1.9557812500000002 A*h | 33.0 V | 33.0 V | 32.5 V | 32.5 V | 28.5 V | 28.5 V | -0.05 V/C | -0.05 V/C | 0.15000000000000002 |
| 12 | 1 | 1442 | 57 | 7 | WLM | 1.9548437500000002 A*h | 33.0 V | 33.0 V | 32.5 V | 32.5 V | 28.5 V | 28.5 V | -0.05 V/C | -0.05 V/C | 0.15000000000000002 |
| 13 | 1 | 1442 | 58 | 7 | WLM | 1.9539062500000002 A*h | 33.0 V | 33.0 V | 32.5 V | 32.5 V | 28.5 V | 28.5 V | -0.05 V/C | -0.05 V/C | 0.15000000000000002 |
| 14 | 1 | 1442 | 59 | 7 | WLM | 1.9529687500000001 A*h | 33.0 V | 33.0 V | 32.5 V | 32.5 V | 28.5 V | 28.5 V | -0.05 V/C | -0.05 V/C | 0.15000000000000002 |
| 15 | 1 | 1442 | 60 | 7 | WLM | 1.95203125 A*h | 33.0 V | 33.0 V | 32.5 V | 32.5 V | 28.5 V | 28.5 V | -0.05 V/C | -0.05 V/C | 0.15000000000000002 |
| 16 | 1 | 1442 | 61 | 7 | WLM | 1.95109375 A*h | 33.0 V | 33.0 V | 32.5 V | 32.5 V | 28.5 V | 28.5 V | -0.05 V/C | -0.05 V/C | 0.15000000000000002 |
| 17 | 1 | 1442 | 62 | 7 | WLM | 1.9501562500000003 A*h | 33.0 V | 33.0 V | 32.5 V | 32.5 V | 28.5 V | 28.5 V | -0.05 V/C | -0.05 V/C | 0.15000000000000002 |
| 18 | 1 | 1442 | 63 | 7 | WLM | 1.9492187500000002 A*h | 33.0 V | 33.0 V | 32.5 V | 32.5 V | 28.5 V | 28.5 V | -0.05 V/C | -0.05 V/C | 0.15000000000000002 |
| 19 | 1 | 1442 | 64 | 7 | WLM | 1.9482812500000002 A*h | 33.0 V | 33.0 V | 32.5 V | 32.5 V | 28.5 V | 28.5 V | -0.05 V/C | -0.05 V/C | 0.15000000000000002 |
| 20 | 1 | 1442 | 65 | 7 | WLM | 1.9473437500000002 A*h | 33.0 V | 33.0 V | 32.5 V | 32.5 V | 28.5 V | 28.5 V | -0.05 V/C | -0.05 V/C | 0.15000000000000002 |
| 21 | 1 | 1442 | 66 | 7 | WLM | 1.9462500000000003 A*h | 33.0 V | 33.0 V | 32.5 V | 32.5 V | 28.5 V | 28.5 V | -0.05 V/C | -0.05 V/C | 0.15000000000000002 |
| 22 | 1 | 1442 | 67 | 7 | WLM | 1.94546875 A*h | 33.0 V | 33.0 V | 32.5 V | 32.5 V | 28.5 V | 28.5 V | -0.05 V/C | -0.05 V/C | 0.15000000000000002 |
| 23 | 1 | 1442 | 68 | 7 | WLM | 1.9443750000000002 A*h | 33.0 V | 33.0 V | 32.5 V | 32.5 V | 28.5 V | 28.5 V | -0.05 V/C | -0.05 V/C | 0.15000000000000002 |

| | A | B | C | D | E | F | G | H | I | J | K | L | M | N |
|---|---|---|---|---|---|---|---|---|---|---|---|---|---|---|
| 1 | SUBSYSTEMID | Session | NOM_KADR | VERPR | REJIM_RAB | RHH | USL_N_UFD | USL_N_UFN | USL_N_UB1D | USL N UB1N | USL_N_UB2C | USL_N_UB2N | USL_N_ATD | USL_N_ATN |
| 102882 | 1 | 2021 | 237 | 7 | WLM | 1.09609375 A*h | 33.0 V | 33.0 V | 32.5 V | 32.5 V | 28.5 V | 28.5 V | -0.05 V/C | -0.05 V/C |
| 102883 | 1 | 2021 | 238 | 7 | WLM | 1.0967187500000002 A*h | 33.0 V | 33.0 V | 32.5 V | 32.5 V | 28.5 V | 28.5 V | -0.05 V/C | -0.05 V/C |
| 102884 | 1 | 2021 | 239 | 7 | WLM | 1.0971875 A*h | 33.0 V | 33.0 V | 32.5 V | 32.5 V | 28.5 V | 28.5 V | -0.05 V/C | -0.05 V/C |
| 102885 | 1 | 2021 | 240 | 7 | WLM | 1.09765625 A*h | 33.0 V | 33.0 V | 32.5 V | 32.5 V | 28.5 V | 28.5 V | -0.05 V/C | -0.05 V/C |
| 102886 | 1 | 2021 | 241 | 7 | WLM | 1.098125 A*h | 33.0 V | 33.0 V | 32.5 V | 32.5 V | 28.5 V | 28.5 V | -0.05 V/C | -0.05 V/C |
| 102887 | 1 | 2021 | 242 | 7 | WLM | 1.0987500000000001 A*h | 33.0 V | 33.0 V | 32.5 V | 32.5 V | 28.5 V | 28.5 V | -0.05 V/C | -0.05 V/C |
| 102888 | 1 | 2021 | 243 | 7 | WLM | 1.0992187500000001 A*h | 33.0 V | 33.0 V | 32.5 V | 32.5 V | 28.5 V | 28.5 V | -0.05 V/C | -0.05 V/C |
| 102889 | 1 | 2021 | 244 | 7 | WLM | 1.0996875000000002 A*h | 33.0 V | 33.0 V | 32.5 V | 32.5 V | 28.5 V | 28.5 V | -0.05 V/C | -0.05 V/C |
| 102890 | 1 | 2021 | 245 | 7 | WLM | 1.1001562500000002 A*h | 33.0 V | 33.0 V | 32.5 V | 32.5 V | 28.5 V | 28.5 V | -0.05 V/C | -0.05 V/C |
| 102891 | 1 | 2021 | 246 | 7 | WLM | 1.10078125 A*h | 33.0 V | 33.0 V | 32.5 V | 32.5 V | 28.5 V | 28.5 V | -0.05 V/C | -0.05 V/C |
| 102892 | 1 | 2021 | 247 | 7 | WLM | 1.10125 A*h | 33.0 V | 33.0 V | 32.5 V | 32.5 V | 28.5 V | 28.5 V | -0.05 V/C | -0.05 V/C |
| 102893 | 1 | 2021 | 248 | 7 | WLM | 1.10171875 A*h | 33.0 V | 33.0 V | 32.5 V | 32.5 V | 28.5 V | 28.5 V | -0.05 V/C | -0.05 V/C |
| 102894 | 1 | 2021 | 249 | 7 | WLM | 1.1023437500000002 A*h | 33.0 V | 33.0 V | 32.5 V | 32.5 V | 28.5 V | 28.5 V | -0.05 V/C | -0.05 V/C |
| 102895 | 1 | 2021 | 250 | 7 | WLM | 1.1028125000000002 A*h | 33.0 V | 33.0 V | 32.5 V | 32.5 V | 28.5 V | 28.5 V | -0.05 V/C | -0.05 V/C |
| 102896 | 1 | 2021 | 251 | 7 | WLM | 1.10328125 A*h | 33.0 V | 33.0 V | 32.5 V | 32.5 V | 28.5 V | 28.5 V | -0.05 V/C | -0.05 V/C |
| 102897 | 1 | 2021 | 252 | 7 | WLM | 1.10375 A*h | 33.0 V | 33.0 V | 32.5 V | 32.5 V | 28.5 V | 28.5 V | -0.05 V/C | -0.05 V/C |
| 102898 | 1 | 2021 | 253 | 7 | WLM | 1.104375 A*h | 33.0 V | 33.0 V | 32.5 V | 32.5 V | 28.5 V | 28.5 V | -0.05 V/C | -0.05 V/C |
| 102899 | 1 | 2021 | 254 | 7 | WLM | 1.1048437500000001 A*h | 33.0 V | 33.0 V | 32.5 V | 32.5 V | 28.5 V | 28.5 V | -0.05 V/C | -0.05 V/C |
| 102900 | 1 | 2021 | 255 | 7 | WLM | 1.1053125000000001 A*h | 33.0 V | 33.0 V | 32.5 V | 32.5 V | 28.5 V | 28.5 V | -0.05 V/C | -0.05 V/C |
| 102901 | 1 | 2021 | 0 | 7 | WLM | 1.1059375 A*h | 33.0 V | 33.0 V | 32.5 V | 32.5 V | 28.5 V | 28.5 V | -0.05 V/C | -0.05 V/C |
| 102902 | 1 | 2021 | 1 | 7 | WLM | 1.10640625 A*h | 33.0 V | 33.0 V | 32.5 V | 32.5 V | 28.5 V | 28.5 V | -0.05 V/C | -0.05 V/C |

**Fig. 9** Screenshots of the unpacked packet of EGYSAT1

interface results, we conclude that this unpacking system is flexible which make it a generic system compatible with other structure of satellite system.

**Acknowledgements** This work is supported by Egypt Knowledge and Technology Alliance (E-KTA) for Space Science "TEDSAT1", which is supported by The Academy of Scientific Research & Technology (ASRT), and coordinated by National Authority for Remote Sensing and Space Sciences (NARSS).

## References

1. L. Zhou, A. Junshe, Design of a payload data handling system for satellites, in *Third International Conference on Instrumentation, Measurement, Computer, Communication and Control (IMCCC)* (IEEE, Shenyang, China, 2013)
2. A. Nicolai, S. Roemer, S. Eckert, The TET satellite bus—future mission capabilities, in *Aerospace Conference* (IEEE, Big Sky, MT, USA, 2014)
3. B. Anyaegbunam, Design elements of satellite telemetry, tracking and control subsystems for the proposed Nigerian made satellite. Int. J. Eng. Sci. Invention **3**(1), 5–13 (2014)
4. P.K. Udaniya, G. Sharma, L. Tharani, Application of MIMO system for telemetry, tracking command and monitoring subsystem to control the satellite, in *International Conference on Computing, Communication and Automation (ICCCA2016)* (IEEE, Greater Noida, India, 2016)
5. T. Yairi, Y. Kawahara, R. Fujimaki, Y. Sato and K. Machida, Telemetry-mining: a machine learning approach to anomaly detection and fault diagnosis for space systems, in *2nd IEEE International Conference on Space Mission Challenges for Information Technology* (IEEE, CA, USA, 2006)
6. R. Fujimaki, T. Yairi, K. Machida, Adaptive limit-checking for spacecraft using relevance vector autoregressive model, in *8th International Symposium on Artificial Intelligence, Robotics and Automation in Space—iSAIRAS, ESA SP-603*, Munich, Germany (2005)
7. M.A. Hearst, S.T. Dumais, E. Osuna, J. Platt, B. Scholkopf, Support vector machines. IEEE Intell. Syst. Appl. **13**(4), 18–28 (1998)
8. W. Qiang, D. Xuan, Analysis of support vector machine classification. Comput. Anal. Appl. **8**(2), 99–119 (2006)
9. S. Bhaskar, K. Goebel, S. Poll, J. Christophersen, Prognostics methods for battery health monitoring using a Bayesian framework. IEEE Trans. Instrum. Measurement,, Vol. 58, No. 2, pp. 291–296, 2009
10. S. Bhaskar, K. Goebel, J. Christophersen.: Comparison of prognostic algorithms for estimating remaining useful life of batteries, Transactions of the Institute of Measurement and Control, Vol. 31, No. 4, pp. 293–308, 2009
11. Y. Song, D. Liuy, Y. Hou, J. Yu, Y. Peng, Satellite lithium-ion battery remaining useful life estimation with an iterative updated RVM fused with the KF algorithm. Chin. J. Aeronaut. **31**, 31–40 (2018)
12. Y. Jinsong, M. Baohua, T. Diyin, L. Hao, W. Jiuqing, Remaining useful life prediction for lithium-ion batteries using a quantum particle swarm optimization-based particle filter. Quality Engineering Journal, Special Issue on Reliability Engineering, Vol. 29, pp. 536–546, 2017
13. H. Thiago, R. Donato, M.G. Quiles, Machine learning systems based on xgBoost and MLP neural network applied in satellite lithium-ion battery sets impedance estimation. Adv. Comput. Intell. Int. J. **5**(1), 1–20 (2018)
14. Y. Rottenstreich, A. Tversky, Unpacking, repacking, and anchoring: advances in support theory. Psychol. Rev. **104**(2), 406–415 (1997)
15. H.R. Glahn, *On the packing of grid point data for efficient transmission. TDL Office Note 92–11* (National Weather Service, NOAA, U.S. Department of Commerce, 1992)

# Multiscale Satellite Image Classification Using Deep Learning Approach

**Noureldin Laban, Bassam Abdellatif, Hala M. Ebied, Howida A. Shedeed and Mohamed F. Tolba**

**Abstract** Image classification has been acquiring special importance in the practical applications of remote sensing. This is done with the extraordinary rise of spatial and spectral resolution of satellite imaging sensors. Also it comes from the daily increase of remote sensing databases. Deep learning approaches, especially Convolutional Neural Networks (CNNs) techniques, have been recently outperforming other state-of-the-art classification approaches in various domains. In this chapter, we propose an enhanced technique for classification of satellite images using CNNs. There are two characteristics of satellite images that make performance issue very crucial; first, high information content within the satellite image, and secondly, high computational requisites involved by CNNs. The improvement technique is built on an effective selection of suitable image scale. As this scale achieves a respectively high classification accuracy alongside a minimal computational use. We conduct our proposed technique using three state-of-the-art datasets: WHU-RS Dataset, UCMerced Land Use Dataset, and Brazilian Coffee Scenes Dataset. The proposed technique results in enhancing the accuracy performance, instead of using the original scale directly.

## 1 Introduction

The recent growth of remote sensing data, from various satellites, has led to an extremely large interest in advanced remote sensing data mining techniques to computerize the extraction of remote sensing information from their massive datasets [1]. A variety of remote sensing systems specification is distributed among a different satellite operators and manufacturers. There are different products and their appli-

N. Laban (✉) · B. Abdellatif
Data Reception and Analysis Division, National Authority for Remote, Sensing and Space Science, Cairo, Egypt
e-mail: nourlaban@gmail.com

H. M. Ebied · H. A. Shedeed · M. F. Tolba
Faculty of Computer and Information Sciences, Ain Shams University, Cairo, Egypt



A. E. Hassanien et al. (eds.), *Machine Learning and Data Mining in Aerospace Technology*, Studies in Computational Intelligence 836,
https://doi.org/10.1007/978-3-030-20212-5_9

cations for remote sensing data. The high-resolution satellite images have several applications. These applications include mapping, planning (engineering, natural resources, urban, infrastructure), change detection, land-use, tourism, crop management, military, and environmental monitoring. Also, they allow us to solve various problems such as monitoring state of the environment and influence of anthropogenic factors, detecting contaminated territories, unauthorized buildings, estimating the state of forest plantation, carrying out operational monitoring of land resources, city building and other [2].

Classification of the visual data is one of the most important steps in almost any computer vision problem, including in the remote sensing domain. Classification of high-resolution satellite images has many new research topics in the remote sensing field. Advanced methodologies have significantly contributed to the solution of the VHR classification problem in the last years [3]. Classification process has two main questions; first how to identify a target features; second, how to use this identification to catch the new one. Since manual identification of these features is not practical in most cases, during years, substantial efforts have been dedicated for developing automatic and discriminating visual feature descriptors. Matching between old and new objects is the essence of machine learning techniques [4].

The remote sensing image classification process has several problems as land-cover and land-use maps are necessary for multi-temporal researches and produce useful information to other processes [5]. The important challenges are as follow:

1. Complex statistical characteristics of satellite images:
   The statistical properties of the satellite images have many difficulties for automatic classifiers. The extraction of information from these images is very challenging, as they have a very high spatial and spectral redundancy and collinearity, a high dimensionality of pixels, a specific noise and uncertainty sources observed, and their potentially nonlinear nature. So, we have to concentrate on the spatial and spectral redundancy also we suggest that the obtained value may be best represented in sparse representation spaces.
2. High-computational requirements:
   Remote sensing data are considered a good example of Big Data. Satellite sensors obtain a huge number of images with different characteristics that have various spectral, spatial, angular, and temporal resolutions. There is an ever-increasing amount of data gathered with current and upcoming Earth Observation (EO) satellite missions, from Multi-Spectral Scanner (MS) sensors such as Landsat-8 to Very High Resolution (VHR) sensors such as WorldView-III, the super spectral Sentinel-2 and Sentinel-3 missions, as well as the planned Environmental Mapping and Analysis Program, Hyper-spectral Infrared Imager, and the European Space Agency's candidate Fluorescence Explorer imaging spectrometer missions. This data stream requires computationally efficient classification techniques.

The kNN classifier is one of the most simple and traditional classification techniques and has been extensively used for geospatial object detection and image classification [6]. It firstly uses labeled training samples as a reference set. Then, we identify a subset of k training samples that are the closest to test sample. Finally, we label it with the class that arrives most frequently with this k subset. The most attractive property of the kNN classifier is the simplicity of the learning rule and the tuning of only one free parameter. Although the selection of the neighbor size k is a challenging problem because different k values will result in different performances. If k is huge, the neighbors searching will take large time whereas a small value of k may decrease the prediction accuracy.

Random Forest (RF) is an ensemble classifier composed of hundreds of decision-tree-based models. RF trains k decision trees for k training subsets sampled randomly with replacement from the original sample set. The final classification decision is voted by all classification trees [7]. Random forest has the advantages of automatic balance of error and automatic selection of features. The algorithm is easy to parallelize, so it has very excellent performance in dealing with large-scale imbalanced data classification. The main disadvantage is that it stores the intermediate results to the disk which is insufficient for the efficiency requirement of real-time query [8].

Multilayer Perceptron (MLP) is a neural network which just maps the input features into output through one or more layers between the input and output layers. MLP is considered to be a fully connected directed graph that includes different layers of nodes. Each node is a neuron (i.e. processing unit) which has a nonlinear activation function. The training of the network is performed using a supervised learning technique called back propagation. Each node in one layer is connected to every node in the other layer with a certain weight. The learning is performed by changing these connection weights after processing each piece of data, based on the amount of error in the output compared to the expected result [9].

Support Vector Machine (SVM) classifier is based on statistical learning theory using a non-parametric supervised classification. The SVM training algorithm is designed to get the optimal hyperplanes that differentiate between classes with a minimum error using a transformation technique that maps the training data into higher dimension space. The optimal hyperplanes are designed using training samples found at the boundaries of class distribution in a feature space. The hyperplane of maximum margin is defined using the support vectors selected from training samples where the other training samples are ignored as they do not have any role to calculate hyperplane locations. Consequently, the competitive advantage of SVM is possibility to achieve high classification accuracy using a small number of training samples [10].

Deep Learning is one of the machine learning techniques. It is built on learning different levels of representations. These levels are corresponding to a hierarchical structure of features as higher-level features are denoted using lower-level ones, and the same lower level features can help to denote of many higher-level features. Deep learning is one of learning representations methods. An observation can be represented in different ways, but some representations make it easier to learn from examples. Research in this area tries to define what makes better representations and how to learn them [11].

Deep learning algorithms especially Convolutional Neural Networks (CNNs) have recently used in a wide scope of computer vision applications due to their excellent feature representation in [12]. This comes from the powerful ability of CNNs to automatically detect the correlated contextual features in image classification problems. CNNs consist of successive layers of trained convolution filters. These filters learn hierarchical contextual image features, which are the common format of deep learning networks. The CNN feature built using neural networks of deep architecture. So, the created features are directly generated representations from raw image pixels. Thus, the load of feature selection has been transmitted to the network configuration itself [13]. So, the multiple layers and neurons are accountable for the composing of nonlinear processing units. As these Layers discover adaptable feature representations in style-like of hierarchy. Low-level features are trained in the first layers, and the high-level features are trained in the deeper ones according to the data itself. Thus, the network learns the features of different levels which lead to robust classifiers [14].

Dealing with remote sensing data, big data, and convolutional neural networks, computationally demanding in the real time is a really performance challenge [3, 15, 16]. Although there are many advances in hardware, the optimization problem is still a critical issue. In the last few years, there are many remote sensing datasets that have been proposed as benchmarks e.g. UCMerced land-use [4], RS19 dataset [17], Brazilian Coffee Scenes [18]… etc. The recent CNNs implementations [5, 14, 18–22] have achieved very high accuracy, exceeding 96%, using aforementioned state-of-the-art datasets. The challenge now is how to improve the performance in terms of memory size and processing time without affecting the recognition accuracy.

Many works on using CNNs for satellite imagery emerged in the recent five years. The important key feature of algorithms that used CNNs is that they do not require previous feature extraction, thus resulting in increasing generalization capabilities [23]. CNNs have achieved a better performance in many problems. Recently, CNNs have been shown to be successful in object recognition [24], object detection [25], scene parsing [26] and scene classification [27]. Deep learning has also very powerful contribution within remote sensing special data as Hyperspectral (HSI) image [28, 29], Synthetic Aperture Radar (SAR) images [9, 30] and Light Detection and Ranging (LiDAR) images [31].

The remainder of this chapter is organized as follows. Section 2 presents the related work. Section 3 presents some concepts related to deep CNNs and our proposed methodology. The hardware configuration, datasets, and experimental results are presented in Sect. 4. Finally, Sect. 5 concludes the chapter.

## 2 Related Work

Deep Learning algorithms get computationally-expensive during processing of very high-dimension data such as satellite images. This is likely because of the slow learning procedure associated with an increased number of the structured layered hierarchy of learning data. This structure includes abstractions and representations from a lower-level layer to a higher-level layer [32]. The implementation of deep learning techniques for satellite image classification has become an active research topic in the remote sensing community. It is mainly encouraged by the recent availability of high spatial and spectral resolutions data acquired by the new generation of satellites. These techniques are utilized in all applications of satellite image classification. The following work overview deals with the issue of information extraction and data representation using different techniques of deep learning and Convolutional Neural Networks.

Nogueira et al. [14] presented an improvement for Spatial Feature Representation from Aerial scenes by using Convolutional Networks. A strategy for hyperspectral image classification is proposed in [33], where attribute filtered images stacked and provided as an input to convolutional neural networks. Basaeed et al. [34] proposed a region segmentation technique for remote sensing images using a boosted committee of Convolutional Neural Networks (CNNs) coupled with inter-band and intra-band fusion. In [35], the Extreme Learning Machines (ELMs) as a stack of supervised autoencoders are used for synthesizing deep neural networks.

Marmanis et al. [3] presented the ImageNet pretrained networks to deal with the limited-data problem in an end-to-end processing scheme. Zhang et al. [36] developed the hierarchical discriminative feature learning algorithm for hyperspectral image classification, which is a deformation of the spatial-pyramid-matching model. Chan et al. [37] employed the PCA to learn the multistage filter banks which followed by simple binary hashing and block histograms for indexing and pooling. Kussul et al. [38] presented a multilevel architecture to target the land cover and crop type classification from multi-temporal multi-source satellite imagery.

Yao et al. [26] have proposed stacked discriminative sparse autoencoder to learn high-level features on an auxiliary satellite image data set then transferred the learned high-level features to semantic annotation to do their classification. Mei et al. [39] have used a five-layer CNN to learn features of hyperspectral images for classification using advances in deep learning area, such as batch normalization, dropout, and Parametric Rectified Linear Unit (PReLU) activation function. Ferreira et al. [40] introduce a boosting-based technique for classification of regions in regions of interest (RSIs) that manages to encode features extracted from different spectral and spatial domains.

Ikasari1 et al. [41] introduce a fast classification of paddy growth stages using multiple regularizations learning on Deep Neural Networks and 1-D Convolutional Neural Networks using LANDSAT 8 image data obtained from multi-sensor remote sensing image. Lv et al. [42] extend the Local Receptive Field(LRF)-based Extreme Learning Machine (ELM) method to a hierarchical model for hyperspectral image classification. Zhao et al. [43] propose Discriminant Deep Belief Network (DisDBN) approach to learning high-level features for SAR image classification in which the discriminant features are learned by combining ensemble learning with a deep belief network in an unsupervised manner. Zou et al. [44] propose deep-learning-based feature-selection method which base on selecting features that are more reconstructible as the discriminative features as features with smaller reconstruction errors would hold image representation.

Li et al. [45] propose a pixel-pair method to significantly increase the number of training samples to enhance CNN classification. For the training procedure, paired samples are fed with new labels into deep CNN. neighboring pixel-pairs constructed using its surroundings and classified by the trained CNN for each testing pixel, the final label is then determined via a voting strategy. Bentes et al. [46] present workflow for SAR maritime targets detection and classification on TerraSAR-X high-resolution image using multiple input resolution CNN model. Zhou et al. [47] investigate extraction of deep feature representations based on convolutional neural networks (CNN) for high-resolution remote sensing image retrieval using two schemes; the first scheme, the deep features are extracted from the fully-connected and convolutional layers of the pre-trained CNN models and the second scheme, a CNN architecture based on conventional convolution layers and a three-layer perceptron is proposed.

Liu et al. [48] explore DCNN with Spatial Pyramid Pooling (SPP-net) by warping the original satellite image into multiple different scales. Then images in each scale are used to train a Deep Convolutional Neural Network (DCNN). They accelerate the training process using different SPP-nets which have the same number of parameters. Wang et al. [49] introduce a self-learning framework for auto registration of satellite images by learning the mapping function using images and their transformed copies. They apply a transfer learning to reduce the huge computation cost in the training stage. Nai-wen et al. [50] propose an extraction method for cultivated land information based on Deep Convolutional Neural Network and Transfer Learning (DTCLE) using linear features and transfer learning mechanisms.

Volpi et al. [21] propose a CNN-based system depending on down sampling followed by up sampling architecture. It uses convolutions to learn a simple spatial map of high-level representations. Then, uses deconvolutions to upsample them back to the original resolution. So, the CNN can label every pixel at its original resolution of the image, and this leads to increase its effectiveness at inference time. Sokoli et al. [51] study generalization error of deep neural networks via their classification margin and introduce an approach based on the Jacobian matrix of a deep neural network.

## 3 Convolutional Neural Networks

Convolutional Neural Networks (CNNs) is a specialized type of neural network used for data processing that has a familiar topology. It can process images easily as a 2-D grid of pixels [52].

1. Biological Inspiration of CNNs
   A convolutional network layer is inspired by properties of the primary visual cortex to catch images. The primary visual cortex is the first region of the brain that performs advanced processing of visual input. The visual input is the images which are formed through light arriving in the eye and stimulating the photore-

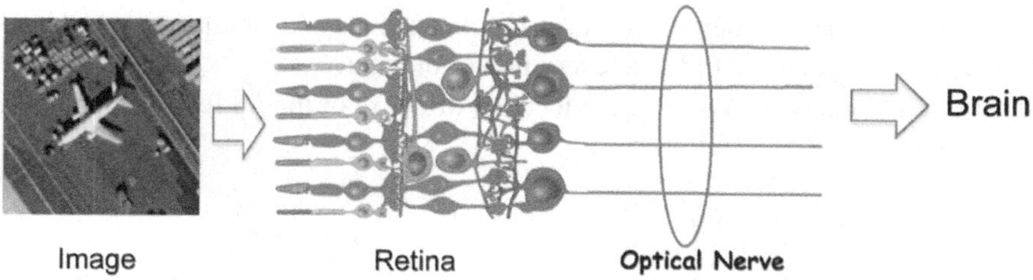


**Fig. 1** Image processing on retina

ceptors in the retina. Neurons in the retina perform concussive convolution like operations to the image until it makes its way to the cortex to be represented. As image light falls on retinal photo-receptor neurons [53], each neuron process a part of the image. This is shown in Fig. 1.

2. Convolutional Neural Network Model
   The learning operation as shown in Fig. 2 is a chain of partial convolutions and pooling. This operation is finished by a fully connected layer. Also. It is affected by three considerations that play a key role in the learning operation of a CNN model: parameter sharing, sparse interaction, and equivariant representation.

A complete CNN architecture has two main layers, these layers are a convolution layer and a pooling layer as follows:

Convolution Layer:

Convolution maintains the spatial relationship between image pixels. This is done by learning features using small parts of the image. The Convolution layer's parameters formed by a group of filters. Every filter is a small size, Also, it extends through the full depth of the input image (bands). We convolve each filter across the width and height of the input image. Then, we compute the dot products between the elements of the filter and the input at all pixels. The filter is slid over the full width and height

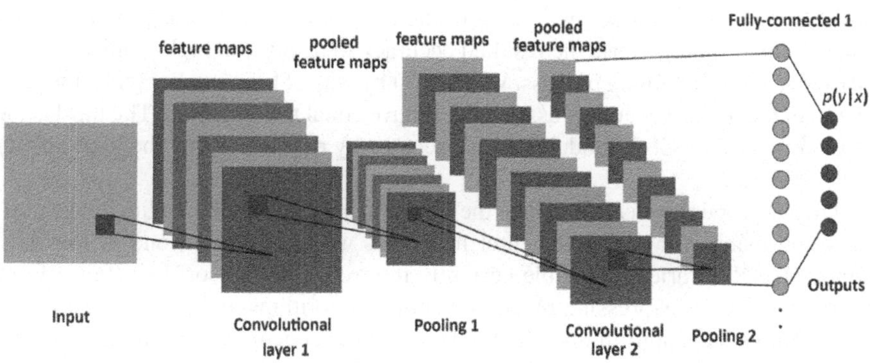


**Fig. 2** Schematic structure of CNNs

of the input image. The output will form a 2-dimensional activation map, which gets from the responses of that filter at every part of the image.

The value of a neuron $v_{\{ij\}}^{\{xy\}}$ at the location of $(x, y)$ of the $j$th feature map in the $i$th layer is expressed as follows:

$$v_{ij}^{xy} = g\left(b_{ij} + \sum_{m}\sum_{p=0}^{P_i-1}\sum_{q=0}^{Q_i-1} w_{ijm}^{pq} v_{(i-1)m}^{(x+p)(y+q)}\right) \tag{1}$$

As m is the feature map in the $(i-1)th$ layer linked to the current ($j$th) feature map, $w_{ijm}^{pq}$ is the weight of location (p, q) linked to the $m$th feature map, $P_i$ and $Q_i$ are the height and the width of the spatial convolution kernel, and $b_{ij}$ is the bias of the $j$th feature map in the $i$th layer [54].

Pooling Layer:

It is used to struggle to overfit by gradually reducing the spatial size of the feature map. this action lead to reduce the number of parameters in the network, which lead to reduce the computational cost. It executed independently on each image band of the input image. this lead to resizes image spatially, usually by using the max operation. A pooling layer can be viewed as a mesh of pooling units diverge s pixels apart, each unit is summing up a surroundings of size z × z centered at its position. If we put s = z, we get classic local pooling as generally implemented in CNNs. If we put s < z, we get an overlapping pooling [24].

## 4 Proposed Methodology

As the proposed approach consists of three levels for overall performance enhancement classification as shown in Fig. 3. The First is the initialization of Convolutional Neural Network (CNN) model. Second is the scale selection process, via this method distinct levels of image scales are proposed with sufficient training time. Using pre-initialized for CNN model, we measure the accuracy of every image scale within enough small time. We use the resulted accuracy to decide the high-quality scale to be used in the full training process. We suggest by way of high-quality scale, the least one with the highest accuracy as all models have equal training time. The third is the full training of the CNN model. This is executed by means of using the high-quality scale selected.

The major purpose is to enhance the overall performance of satellite image classification using convolutional neural networks, we have chosen one of the most famous CNN libraries accessible currently referred to as TensorFlow. TensorFlow is an interface for expressing machine learning algorithms and an implementation for executing such algorithms. The fundamental benefits of TensorFlow are its flexibility and it can be used to express a broad range of algorithms, which include training and inference algorithms for deep neural network models. It is additionally

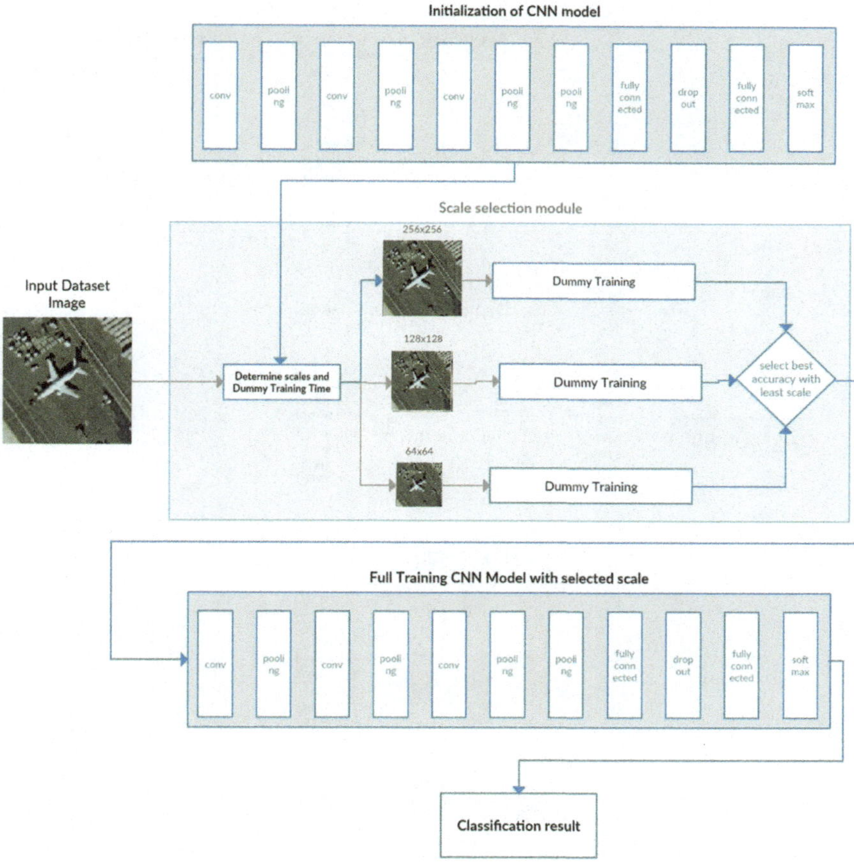


**Fig. 3** Block diagram for performance enhancement classification process

used for conducting research and for deploying machine learning models into industry throughout more than a dozen areas of computer science and different fields. It is included in speech recognition, computer vision, robotics, information retrieval, natural language processing, geographic information extraction, and computational drug discovery [55].

Our CNN model consists of three convolutional layers, three pooling (subsampling) layers, one dropout layer, two fully connected layers and one accuracy (softmax) layer organized as shown in Fig. 4.

Input Data Layer is designed to feed raw image data to the entire network. In this layer, we set the dimensions of data used to train the model. Size of data is determined by width, height and number of bands for every image plus number of images used via the model. The first convolutional layer applies 32 $7 \times 7$ filters to the input layer, with a Rectifier linear ReLU activation function. The second and third convolution layers have 64 $3 \times 3$ filters. Pooling layers, which down-sample

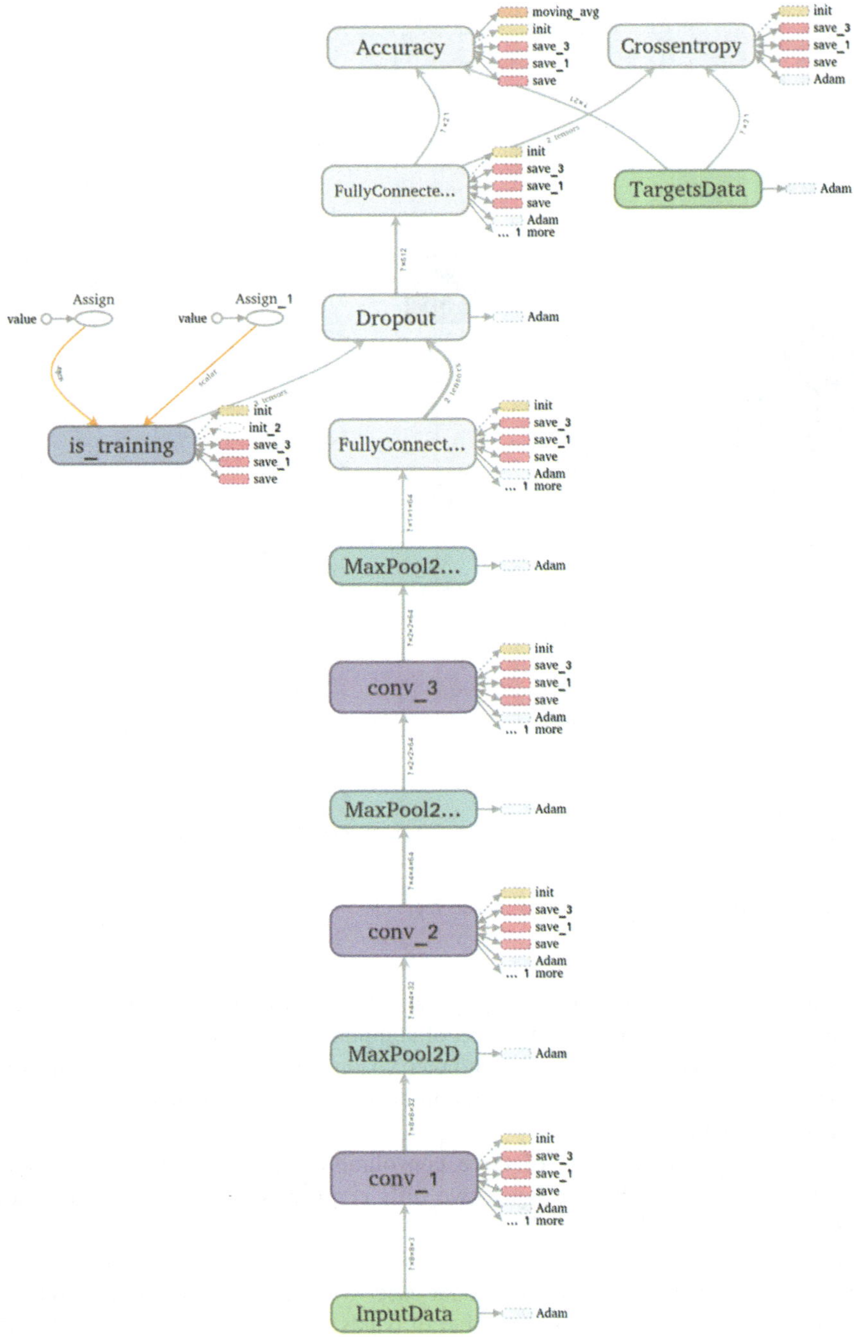


**Fig. 4** Convolution neural networks model

the image data extracted by using the convolutional layers, that lead to a decrease in the dimensionality of the characteristic map which reduces processing time. We used 'max pooling, which extracts sub-regions of the characteristic map (e.g. 2 × 2-pixel tiles), maintains their maximum value, and discards all different values. Fully connected (dense) layers, with 512 neurons, do classification on the features extracted by using the convolutional layers and down-sampled through the pooling layers. In a dense layer, each node in the layer is linked to another node in the preceding layer. Dropout layer helps prevent overfitting. The softmax [56] (accuracy) layer designed to get class with the best probability.

# 5 Experimental Results

In this section, experiments are carried out to decide the effect of altering the image scale on the classification performance. Section 5.1 gives the three remote sensing datasets. Experimental outcomes are introduced. In Sect. 5.2, all the experiments are performed on the same computer with Quad Intel Core i7-6500U CPU @ 2.50 GHz and 16 GB RAM.

## *5.1 Remote Sensing Datasets*

Experiments are performed on three remote sensing datasets. These three datasets are distinctive in spatial and spectral information. The accuracy results are compared with the overall performance of the latest results in these three datasets and with distinctive scales. The three publicly accessible datasets used in our experiments are as follows:

UCMerced Land Use Dataset is acquired from the United States Geological Survey (USGS) National Map: This dataset includes 2100 aerial scene images. Its dimensions are 256 × 256 pixels. they are manually labeled as 21 land use classes. each class has 100 images. As shown in Fig. 5, the inter-class diversity amongst some categories is very small for this dataset, such as "dense residential", "medium residential" and "sparse residential" [4]. There are notably overlapping between classes of the dataset such as the dense residential, medium residential, and sparse residential. As these classes commonly differ in the density of structures. The 21 classes are harbor, intersection, medium density residential, mobile home park, overpass, agricultural, airplane, baseball diamond, beach, buildings, chaparral, dense residential, forest, freeway, golf course, parking lot, river, runway, sparse residential, storage tanks, and tennis courts [14].

WHU-RS Dataset is gathered from Google Earth. This dataset is composed of 950 aerial scene images. Each scene dimension is 600 × 600 pixels. These scenes are uniformly distributed in 19 scene classes. Each class has 50 scenes. Example images for each class are shown in Fig. 6. Images in WHU-RS Dataset and UC

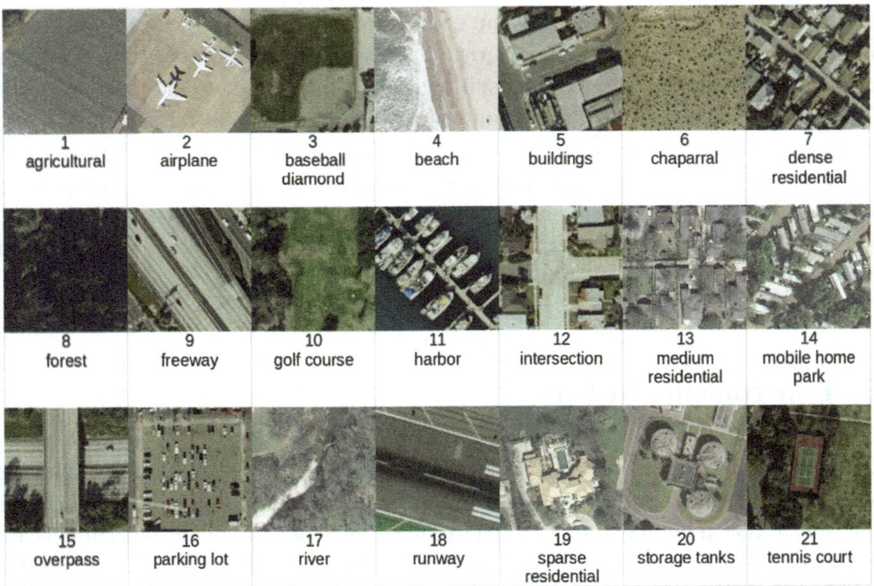


**Fig. 5** UCMerced Land Use Dataset samples

Merced dataset are optical images (RGB color space). They have the same spectral information. However, WHU-RS Dataset images in this dataset contain greater detail information in space. Also, it has a variant of scale and resolution of objects in a large range within the images makes this dataset extra difficult than the UC Merced dataset. This dataset is composed of 19 classes of different scenes, which include desert, farmland, football field, forest, industrial area, meadow, mountain, park, parking, pond, airport, beach, bridge, commercial area, port, railway station, residential area, river and viaduct [57].

Brazilian Coffee Scenes Dataset is taken via the SPOT sensor in the green, red, and near-infrared bands, over four counties in the State of Minas Gerais, Brazil. This dataset is released in 2015. It includes over 50,000 remote sensing images. Each image has 64 × 64 pixels. Also, these images are labeled as coffee (1438) non-coffee (36,577) or mixed (12,989). Figure 7 shows three example images for each of the coffee and non-coffee classes in false colors. Experiments are supplied by a balanced dataset. The experimental dataset used is used as follow: there are 1438 images of both coffee and non-coffee classes are chosen, whilst images of a mixed class are all ignored. The used dataset is very exclusive from the previous two datasets. The specification of the dataset is conducted as follows: tiles with at least 85% of coffee pixels had been assigned to the coffee class where tiles with less than 10% of coffee pixels were assigned to the non-coffee class [57].

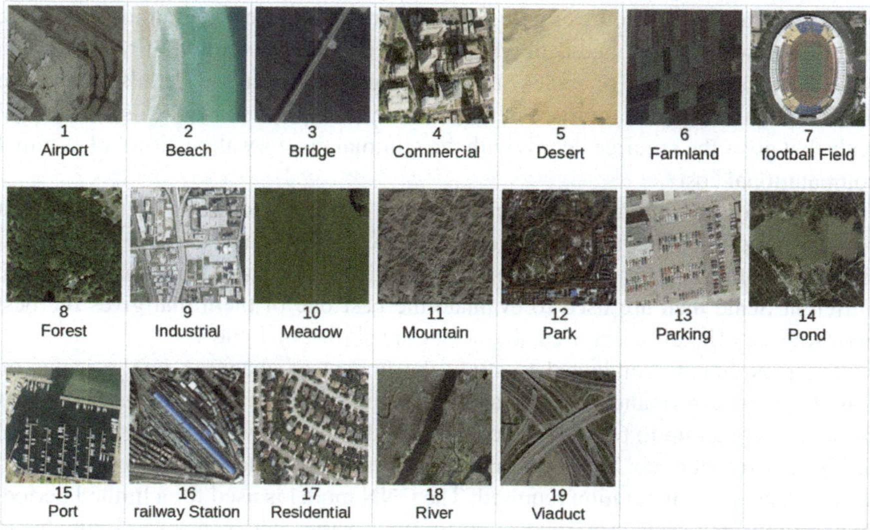


**Fig. 6** WHU-RS Dataset samples

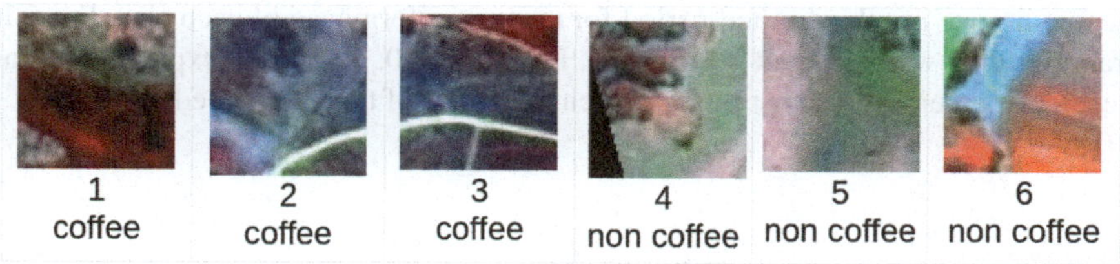


**Fig. 7** Brazilian Coffee Scenes Dataset samples

## 5.2 Results

Analysis of large scale satellite images is challenging and problematic [58, 59]. First challenge, there are a massive amount of details and information with each remote sensing image. As we have different kinds of satellites that thrown terabytes of data every day. Satellite data have wide range of spatial resolutions that may reach to 30 cm. Also, these data have different spectral resolution with may reach 220 band. All these types of data have a tremendous amount of data. Also, these data are achieved in temporary storages increase day by day. This creates huge challenges for data analysis. The second one is the massive amount of Computations needed to process this vast amount of data. Modern models of machine learning algorithms as deep learning need huge computational power. For example, AlexNet, one of the popular deep learning models, have 60 million weights to train and calculated. Each training cycle contain huge number of matrices operations. The third challenge is defining and selecting the suitable scales for processing is the keystone for efficient use of computational resource s with huge amount of data. Scale pyramid as in Fig. 8, is common way to store and deal with satellite images the amount of required data is detected by the level of scale. So, according to the required details, determine the computational effort for these details. Finally, the question is what is the optimal scale that gives the required details with the minimal memory allocation and minimal computational cost?

Our approach configurations are as follow: With respect to datasets, we use a popular remote sensing benchmark datasets (UCMerced Land-use, WHU-RS, And Brazilian Coffee Scenes). Then we use our CNN Model over this benchmark datasets. Different Scale then are used to evaluate the best one of them that gives the best performance. The different used size scales are shown in Table 1.

The experiments conducted to study the impact of changing image scales on classification performance Image scale is resampled from small to large size. It starts from 8 × 8 pixel up to 64 × 64 pixels. This is done by down sampling the original image to a smaller one. This decrease the information content within image which lead to decrease computation required. The CNN model is used for a limited period. According to the scale size, the time of training varies between several minutes to several hours. The optimal scale size with the minimal computational cost should be investigated. So, we determine this limited period to be about 100-unit time. At the beginning, we use this limited period for testing performance of each size. For full training time, we use about 100 epochs. Figures 9, 10, 11 and 12 explain the total accuracy percentage against time for each scale size of the UCMerced dataset.

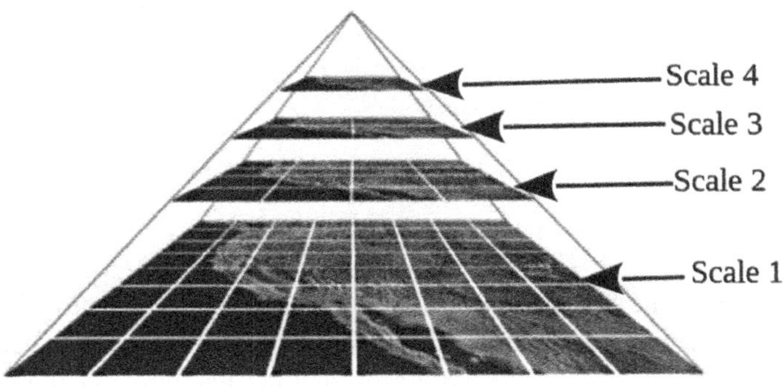


**Fig. 8** Satellite image with scale pyramid

**Table 1** Scale size used with each dataset

| Dataset | Original scale | Scale 1 | Scale 2 | Scale 3 | Scale 4 |
|---|---|---|---|---|---|
| UCMerced Land-use Dataset | 256×256 | 8×8 | 16×16 | 32×32 | 64×64 |
| WHU-RS Dataset | 600×600 | 16×16 | 32×32 | 64×64 | 128×128 |
| Brazilian Coffee Scenes Dataset | 64×64 | 8×8 | 16×16 | 32×32 | 64×64 |

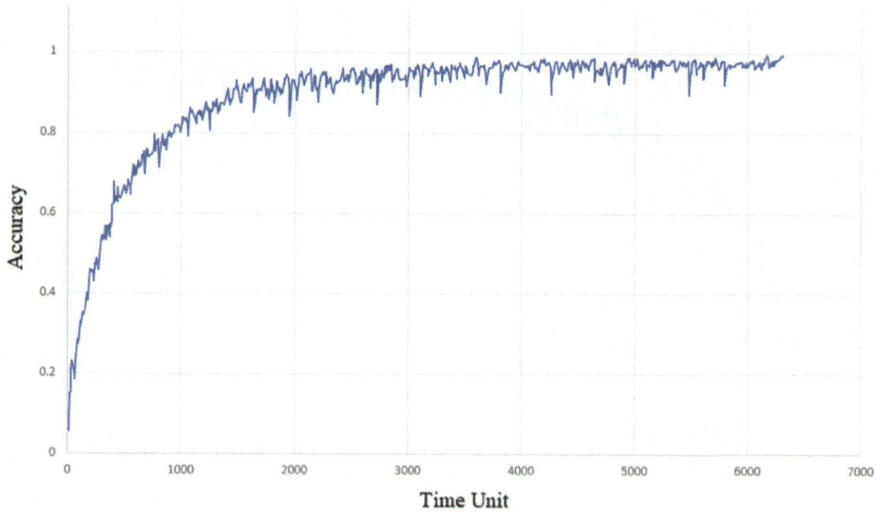


**Fig. 9** 8 × 8 UCMerced training accuracy against time

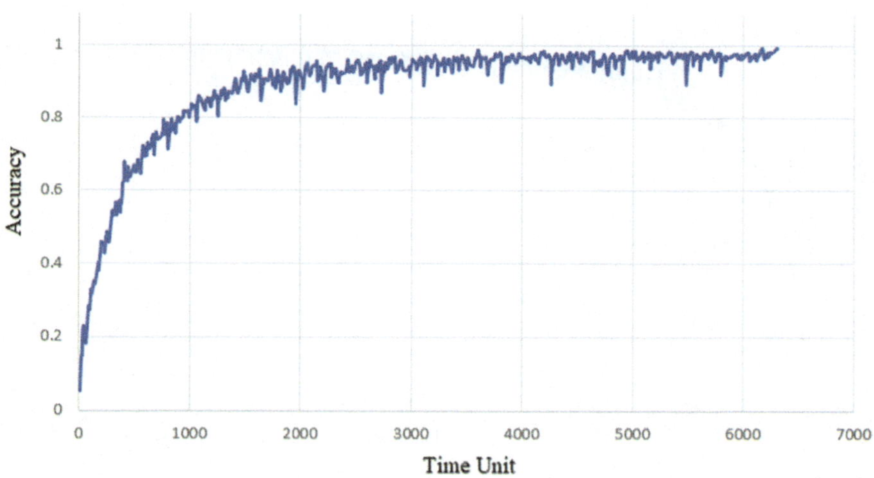


**Fig. 10** 16 × 16 UCMerced training accuracy against time

Experiments are divided to two stages for each training data set. First stage, we have used only a limited time training about 100-unit time (wall time). each data set is scaled to the required scale. Then we train our CNN model with this new scales data. Second stage, full time training process is done for the scale that gives the best accuracy during the first stage. The two stages model gives the best classification accuracy with the minimal training time for the used dataset (Table 2).

With respect to UCMerced Land-use dataset, we have resampled the original size images (256 × 256) to four selected scales as shown in Table 1. Figure 13 shows

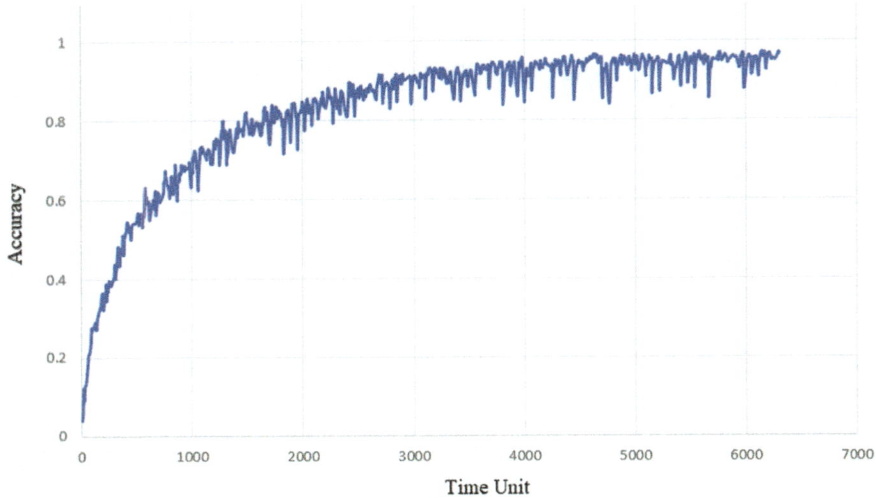


**Fig. 11** 32 × 32 UCMerced training accuracy against time

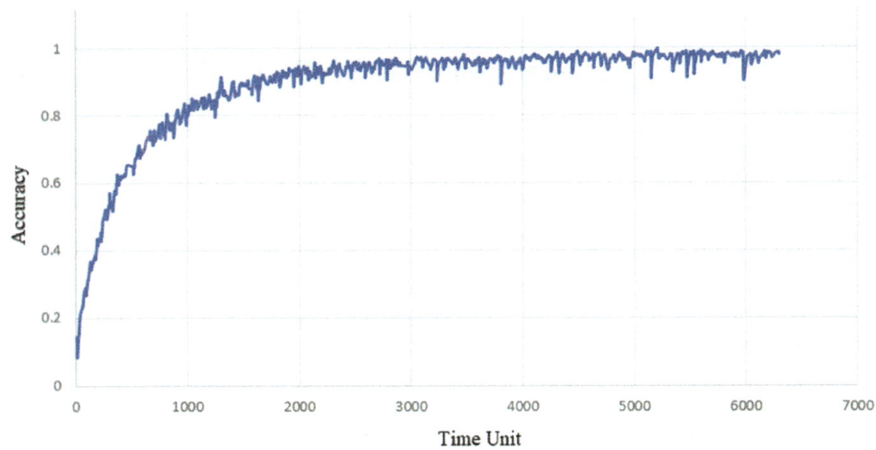


**Fig. 12** 64 × 64 UCMerced training accuracy against time

the relationship between different scale sizes and classification accuracies for each stage. The best accuracy in the first stage is found at scale 2 (16 × 16), as which is the least scale that gives the highest accuracy. With respect to scales 3 and 4, they give high accuracies but with a bigger size. In the same time, the increase in accuracy in the case of scales 3 and 4 is not significant. Also, the results of stage 2 using scale 2 have give also the best overall accuracy.

With respect to WHU-RS dataset, we have also resampled the original size images (600 × 600) to four selected scales as shown in Table 1. Figure 14 shows the relationship between different scale sizes and classification accuracies for each stage. The

**Table 2** Classification accuracy after 100 Unit time and 100 epochs for each dataset

| Dataset | Scale 1 | | Scale 2 | | Scale 3 | | Scale 4 | |
|---|---|---|---|---|---|---|---|---|
| | After 100 time unit | After 100 epoch | After 100 time unit | After 100 epoch | After 100 time unit | After 100 epoch | After 100 time unit | After 100 epoch |
| UCMerced Land-use Dataset | 18.8 | 70.8 | 91.5 | 99.4 | 89.4 | 96.6 | 88.4 | 98 |
| WHU-RS Dataset | 60.7 | 89.7 | 64.9 | 95.9 | 56.4 | 97.7 | 39.7 | 97.4 |
| Brazilian Coffee Scenes Dataset | 78.9 | 94.6 | 88.7 | 96.8 | 88.3 | 98 | 82.6 | 99.3 |

scale that gives the best accuracy during stage 1 is scale 2 (32 × 32). When select scale 2 for stage 2 training is give very high accuracy. The improve in accuracy resulted from using scales 3 and 4 is very small.

Finally, we have resampled Brazilian Coffee Scenes dataset with original size 64 × 64 to for four selected scales as shown in Table 1. Figure 15 shows the relationship between different scale sizes and classification accuracies for each stage. As the data

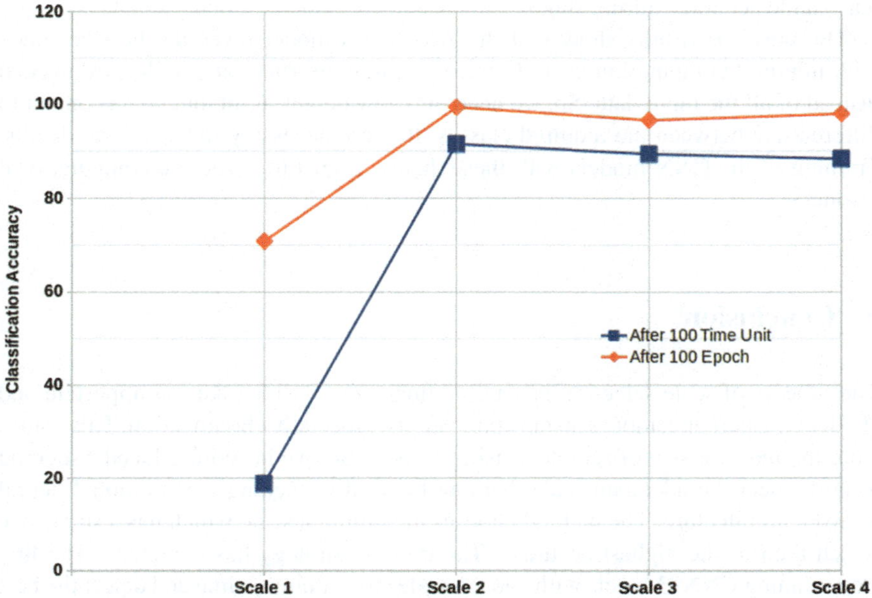


**Fig. 13** UCMerced Land-use Dataset results

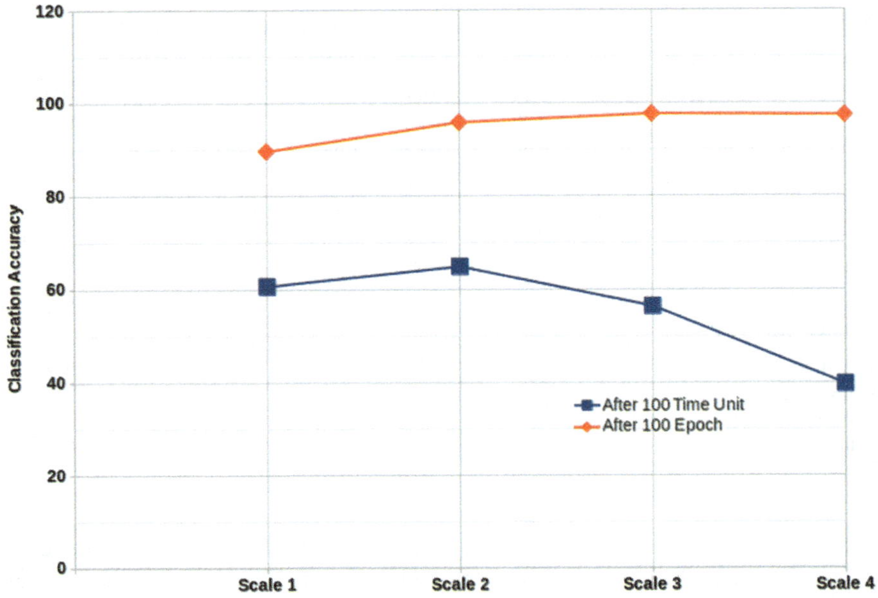


**Fig. 14** WHU-RS Dataset results

set has only two classes (coffee or non-coffee), It give the best accuracies. Although, with respect to the stage 1, scale 2 gives the best accuracy. Scale 2 also gives a high reasonable accuracy during stage.

The previous results show that the two stages model gives the best accuracy with minimal computation cost. Therefore, a few information details. Are needed instead of all the input data. So, we need only few details as an input data content to differentiate between the required classes. It is not necessary to use all the details. Training of the CNN models with these data will lead to waste the computational resources.

## 6 Conclusion

The process of scale selection in satellite image processing takes an important and challenging part on remote sensing applications, especially classification. This comes from the massive size of remote sensing images. So, we have introduced a method to best select the adequate scale that can be used in feeding convolutional neural network architecture. The method chooses the minimal scale which has a small size which records the highest accuracy. The training strategy has two steps. The first step, training CNNs Models with chosen scales for a dummy time and detect the best performance through this time. The second step, after selecting the best scale with

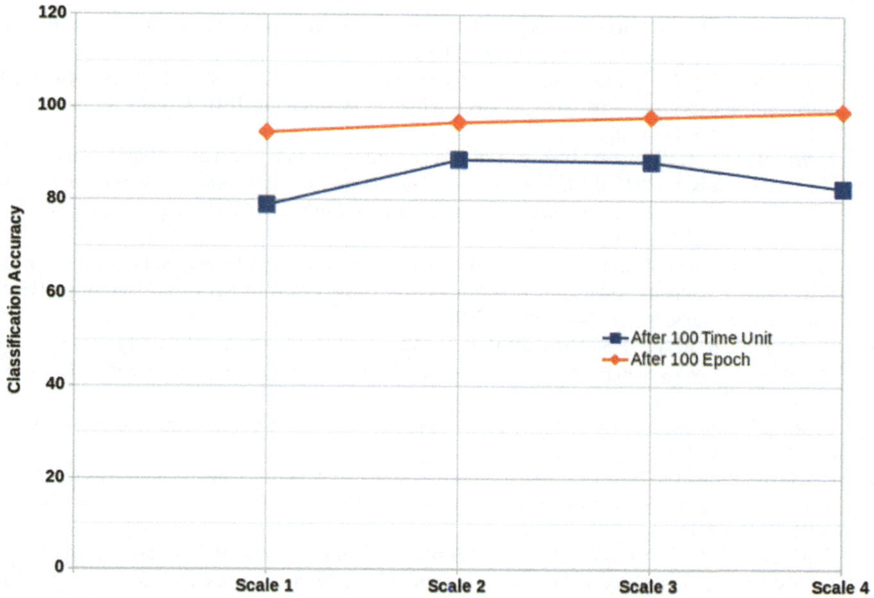


**Fig. 15** Brazilian Coffee Scenes Dataset results

the highest accuracy from the first stage, a complete training process for the CNN model is conducted. Experiments demonstrate that the proposed approach improves performance with respect to the original scale with suitable accuracy. In the future, we think that embedding the adaptive scale mechanism within the CNN architecture instead of using a previous layer may get more accurate results.

## References

1. M. Das, S.K. Ghosh, Deep-STEP: a deep learning approach for spatiotemporal prediction of remote sensing data. IEEE Geosci. Remote Sens. Lett. **13**(12), 1984–1988 (2016)
2. D.M.M. Hordiiuk, V.V.V Hnatushenko, Neural network and local laplace filter methods applied to very high resolution remote sensing imagery in urban damage detection, in *2017 IEEE International Young Scientists Forum on Applied Physics and Engineering (YSF)* (2017), pp. 363–366
3. D. Marmanis, M. Datcu, T. Esch, U. Stilla, S. Member, Deep learning earth observation classification using ImageNet pretrained networks. IEEE Geosci. Remote Sens. Lett. **13**(1), 1–5 (2015)
4. Y. Yang, S. Newsam, Bag-of-visual-words and spatial extensions for land-use classification, in *Proceedings of the 18th SIGSPATIAL International Conference on Advances in Geographic Information Systems* (2010), pp. 270–279
5. A. Romero, C. Gatta, G. Camps-valls, S. Member, Unsupervised deep feature extraction for remote sensing image classification. IEEE Geosci. Remote Sens. Lett. **54**(3), 1–14 (2015)

6. G. Cheng, J. Han, A survey on object detection in optical remote sensing images. ISPRS J. Photogramm. Remote Sens. **117**, 11–28 (2016)
7. Z. Liu, B. Tang, X. He, Q. Qiu, F. Liu, Class-specific random forest with cross-correlation constraints for spectral—spatial hyperspectral image classification. IEEE Geosci. Remote Sens. Lett. **14**(2), 257–261 (2017)
8. Z. Wu, W. Lin, Z. Zhang, A. Wen, L. Lin, An ensemble random forest algorithm for insurance big data analysis, in *2017 IEEE International Conference on Computational Science and Engineering (CSE) and IEEE International Conference on Embedded and Ubiquitous Computing (EUC)*, vol. 5 (2017), pp. 531–536
9. T. L. M. Barreto et al. Classification of detected changes from multitemporal high-res Xband SAR images: intensity and texture descriptors from SuperPixels. IEEE J. Sel. Top. Appl. Earth Obs. Remote Sens. **9**(12), 5436–5448 (2016)
10. B. Zheng, S.W. Myint, P.S. Thenkabail, R.M. Aggarwal, A support vector machine to identify irrigated crop types using time-series Landsat NDVI data. Int. J. Appl. Earth Obs. Geoinformation **34**(1), 103–112 (2015)
11. L. Deng, D. Yu, Deep learning: methods and applications. Found Trends Signal Process. **7**(3–4), pp. 197–387 (2014)
12. G. Cheng, P. Zhou, J. Han, Learning rotation-invariant convolutional neural networks for object detection in VHR optical remote sensing images. IEEE Trans. Geosci. Remote Sens. **54**(99), 7405–7415 (2016)
13. E. Maggiori, Y. Tarabalka, G. Charpiat, P. Alliez, Convolutional neural networks for large-scale remote-sensing image classification. IEEE Trans. Geosci. Remote Sens. **55**(2), 645–657 (2016)
14. K. Nogueira, W.O. Miranda, J.A. Dos Santos, Improving spatial feature representation from aerial scenes by using convolutional networks, in *Brazilian Symposium on Computer Graphics and Image Processing*, vol. 2015, pp. 289–296 (2015)
15. A. Fernández, Á. Gómez, F. Lecumberry, Á. Pardo, I. Ramírez, Pattern recognition in Latin America in the 'big data' era. Pattern Recognit. **48**(4), 1181–1192 (2015)
16. L. Zhou, S. Pan, J. Wang, A.V. Vasilakos, Machine learning on Big Data: opportunities and challenges. Neurocomputing **237**(January), 350–361 (2017)
17. G.-S. Xia et al., Structural high-resolution satellite image indexing, in *ISPRS TC VII Symposium—100 Years ISPRS*, vol. XXXVIII, pp. 298–303 (2010)
18. A.B. Penatti, K. Nogueira, J.A. Santos, O.A.B. Penatti, K. Nogueira, J.A. dos Santos, Do deep features generalize from everyday objects to remote sensing and aerial scenes domains? in *2015 IEEE Conference on Computer Vision and Pattern Recognition Workshops (CVPRW)*, pp. 44–51 (2015)
19. K. Nogueira, O.A.B.O.A.B. Penatti, J.A. Dos Santos, Towards better exploiting convolutional neural networks for remote sensing scene classification. Pattern Recognit. **61**, 539–556 (2016)
20. H. Wu, B. Liu, W. Su, W. Zhang, J. Sun, Deep filter banks for land-use scene classification. IEEE Geosci. Remote Sens. Lett. **13**(12), 1895–1899 (2016)
21. M. Volpi, D. Tuia, Dense semantic labeling of subdecimeter resolution images with convolutional neural networks. IEEE Trans. Geosci. Remote Sens. **55**(2), 881–893 (2016)
22. J. Wang, C. Luo, H. Huang, H. Zhao, S. Wang, Transferring pre-trained deep CNNs for remote scene classification with general features learned from linear PCA network. Remote Sens. **9**(3), 225 (2017)
23. M. Längkvist, A. Kiselev, M. Alirezaie, A. Loutfi, Classification and segmentation of satellite orthoimagery using convolutional neural networks. Remote Sens. **8**(4), 329 (2016)
24. A. Krizhevsky, I. Sutskever, G.E. Hinton, {ImageNet} classification with deep convolutional neural networks. Adv. Neural Inf. Process. Syst. **25**, 1–9 (2012)
25. W. Diao, X. Sun, X. Zheng, F. Dou, H. Wang, K. Fu, Efficient saliency-based object detection in remote sensing images using deep belief networks. IEEE Geosci. Remote Sens. Lett. **13**(2), 137–141 (2016)
26. X. Yao, J. Han, S. Member, G. Cheng, X. Qian, L. Guo, Semantic annotation of high-resolution satellite images via weakly supervised learning. IEEE Trans. Geosci. Remote Sens. **54**(6), 1–12 (2016)

27. C. Farabet, C. Couprie, L. Najman, Y. LeCun, Learning hierarchical features for scene labeling. IEEE Trans. Pattern Anal. Mach. Intell. **35**(8), 1915–1929 (2013)
28. X.Z. Member, S. Li, F.T. Member, K. Qin, S. Hu, S. Liu, Deep learning with grouped features for spatial spectral classification of hyperspectral images. IEEE Geosci. Remote Sens. Lett. **14**(1), 1–5 (2017)
29. D. Tuia, R. Flamary, N. Courty, Multiclass feature learning for hyperspectral image classification: sparse and hierarchical solutions. ISPRS J. Photogramm. Remote Sens. **105**, 272–285 (2015)
30. Y. Zhou, H. Wang, S. Member, F. Xu, S. Member, Y. Jin, Polarimetric SAR image classification using deep convolutional neural networks. IEEE Trans. Geosci. Remote Sens. **13**(12), 1935–1939 (2016)
31. Y. Yu, J. Li, S. Member, H. Guan, C. Wang, Automated detection of three-dimensional cars in mobile laser scanning point clouds using DBM-Hough-Forests. IEEE Trans. Geosci. Remote Sens. **54**(7), 4130–4142 (2016)
32. M.M. Najafabadi, F. Villanustre, T.M. Khoshgoftaar, N. Seliya, R. Wald, E. Muharemagic, Deep learning applications and challenges in big data analytics. J. Big Data **2**(1), 1 (2015)
33. E. Aptoula, M.C. Ozdemir, B. Yanikoglu, Deep learning with attribute profiles for hyperspectral image classification. IEEE Geosci. Remote Sens. Lett. **13**(12), 1970–1974 (2016)
34. E. Basaeed, H. Bhaskar, M. Al-Mualla, Supervised remote sensing image segmentation using boosted convolutional neural networks. Knowl. Based Syst. **99**, 19–27 (2016)
35. M.D. Tissera, M.D. McDonnell, Deep extreme learning machines: supervised autoencoding architecture for classification. Neurocomputing **174**, 42–49 (2016)
36. X. Zhang, Y. Liang, Y. Zheng, J. An, L.C. Jiao, Hierarchical discriminative feature learning for hyperspectral image classification. IEEE Geosci. Remote Sens. Lett. **13**(4), 594–598 (2016)
37. T.H. Chan, K. Jia, S. Gao, J. Lu, Z. Zeng, Y. Ma, PCANet: a simple deep learning baseline for image classification? IEEE Trans. Image Process. **24**(12), 5017–5032 (2015)
38. N. Kussul, M. Lavreniuk, S. Skakun, A. Shelestov, Deep learning classification of land cover and crop types using remote sensing data. IEEE Geosci. Remote Sens. Lett. **14**(5), 778–782 (2017)
39. S. Mei, J. Ji, J. Hou, X. Li, Q. Du, Learning sensor-specific spatial-spectral features of hyperspectral images via convolutional neural networks. IEEE Trans. Geosci. Remote Sens. **55**(8), 4520–4533 (2017)
40. E. Ferreira, A. de A. Araujo, J. A. dos Santos, A boosting-based approach for remote sensing multimodal image classification, in *2016 29th SIBGRAPI Conference on Graphics, Patterns and Images (SIBGRAPI)*, pp. 416–423 (2016)
41. I. H. Ikasari, V. Ayumi, M. I. Fanany, S. Mulyono, Multiple regularizations deep learning for paddy growth stages classification from LANDSAT-8, in *2016 International Conference on Advanced Computer Science and Information Systems (ICACSIS)*, pp. 512–517 (2016)
42. Q. Lv, X. Niu, Y. Dou, J. Xu, Y. Lei, Classification of hyperspectral remote sensing image using hierarchical local-receptive-field-based extreme learning machine. IEEE Geosci. Remote Sens. Lett. **13**(3), 434–438 (2016)
43. Z. Zhao, L. Jiao, J. Zhao, J. Gu, J. Zhao, Discriminant deep belief network for high-resolution SAR image classification. Pattern Recognit. **61**, 686–701 (2016)
44. T. Zhang, Q. Wang, Deep learning based feature selection for remote sensing scene classification. IEEE Geosci. Remote Sens. Lett. **12**(11), 2321–2325 (2015)
45. W. Li, G. Wu, F. Zhang, Q. Du, Hyperspectral image classification using deep pixel-pair features. IEEE Trans. Geosci. Remote Sens. **55**(2), 844–853 (2017)
46. C. Bentes, D. Velotto, B. Tings, Ship classification in TerraSAR-X images with convolutional neural networks. IEEE J. Ocean. Eng. **43**(1), 258–266 (2018)
47. W. Zhou, S. Newsam, C. Li, Z. Shao, Learning low dimensional convolutional neural networks for high-resolution remote sensing image retrieval. Remote Sens. **9**(5), 489 (2017)
48. Q. Liu, R. Hang, H. Song, Z. Li, Learning multiscale deep features for high-resolution satellite image scene classification. IEEE Trans. Geosci. Remote Sens. **56**(1), 117–126 (2018)

49. S. Wang, D. Quan, X. Liang, M. Ning, Y. Guo, L. Jiao, A deep learning framework for remote sensing image registration. ISPRS J. Photogramm. Remote Sens. (2018)
50. H. Lu, X. Fu, C. Liu, L. Li, Y. He, N. Li, Cultivated land information extraction in UAV imagery based on deep convolutional neural network and transfer learning. J. Mt. Sci. **14**(4), 731–741 (2017)
51. J. Sokolic, R. Giryes, G. Sapiro, M.R.D. Rodrigues, Robust large margin deep neural networks. IEEE Trans. Signal Process. **65**(16), 4265–4280 (2017)
52. I. Goodfellow, Y. Bengio, A. Courville, *Deep Learning*. The MIT Press (2016)
53. M.B.A. Djamgoz, S. Vallerga, H.-J. Wagner, Functional organization of the outer retina in aquatic and terrestrial vertebrates: comparative aspects and possible significance to the ecology of vision, in *Adaptive Mechanisms in the Ecology of Vision*, ed. by S.N. Archer, M.B.A. Djamgoz, E.R. Loew, J.C. Partridge, S. Vallerga (Springer, Netherlands, Dordrecht, 1999), pp. 329–382
54. Y. Chen, H. Jiang, C. Li, X. Jia, P. Ghamisi, Deep feature extraction and classification of hyperspectral images based on convolutional neural networks. IEEE Trans. Geosci. Remote Sens. **54**(10), 6232–6251 (2016)
55. M. Abadi, et al., TensorFlow: a system for large-scale machine learning, in *Proceedings of the 12th USENIX Conference on Operating Systems Design and Implementation*, pp. 265–283 (2016)
56. P. Reverdy, N.E. Leonard, Parameter estimation in softmax decision-making models with linear objective functions. IEEE Trans. Autom. Sci. Eng. **13**(1), 54–67 (2015)
57. G. Cheng, J. Han, X. Lu, Remote sensing image scene classification: BENCHMARK and state of the art. **105**(10), 1–19, arXiv:1703.00121 *[cs.CV]* (2001)
58. X. Bian, C. Chen, L. Tian, Q. Du, Fusing local and global features for high-resolution scene classification. IEEE J. Sel. Top. Appl. Earth Obs. Remote Sens. (99), 1–13 (2017)
59. Y. Zhou, J. Li, L. Feng, X. Zhang, X. Hu, Adaptive scale selection for multiscale segmentation of satellite images. IEEE J. Sel. Top. Appl. Earth Obs. Remote Sens. (99), 1–11 (2017)

# Part III
# Security Issues in Telemetry Data

# Security Approaches in Machine Learning for Satellite Communication

**Mamata Rath and Sushruta Mishra**

**Abstract** The emerging technical approach Machine Learning (ML) is apprehensive with the design and growth of algorithms and techniques that allocate computers to “learn”. The major focus of ML research is to extract information from data automatically, by computational and statistical methods. It is thus closely related to data mining and statistics. The power of neural networks stems from their representation capability. In many applications including current discussion of security in satellite communication, feed forward networks are proved to offer the capability of universal function approximation. This chapter thrashes out in details and highlights on important technical issues during machine learning strategies in developing satellite communication systems.

## 1 Introduction

The development and advancement of satellite communication in networking systems require strong and efficient security plans. As a developing innovation, the Internet of Things (IoT) acquires cyber-attacks and dangers from the IT condition in spite of the presence of a layered guarded security instrument. The augmentation of the computerized world to the physical condition of IoT brings inconspicuous attacks that require a novel lightweight and conveyed attack detection system because of their engineering and asset limitations [1, 2]. Compositionally, Fog computing based mobile stations can be utilized to offload security capacities from IoT and the cloud to moderate the asset restriction issues of IoT and versatility bottlenecks of the cloud [3, 4]. This section will further focus on machine learning strategy for better security systems in satellite communication.

M. Rath (✉)
Birla Global University, Bhubaneswar, India
e-mail: mamata.rath200@gmail.com

S. Mishra
KIIT University, Bhubaneswar, India



A. E. Hassanien et al. (eds.), *Machine Learning and Data Mining in Aerospace Technology*, Studies in Computational Intelligence 836,
https://doi.org/10.1007/978-3-030-20212-5_10

While the past "learning by memorization" approach is in some cases helpful, it comes up short on an imperative part of learning frameworks—the capacity to mark concealed email messages. A fruitful student ought to have the capacity to advance from individual guides to more extensive speculation. This is additionally alluded to as inductive thinking or inductive surmising. In the trap bashfulness model displayed already, after the rodents experience a case of a particular kind of sustenance, they apply their frame of mind toward it on new, inconspicuous instances of nourishment of comparative smell and taste [5]. To accomplish speculation in the spam separating errand, the student can examine the recently observed messages, and concentrate a lot of words whose appearance in an email message is demonstrative of spam. At that point, when another email arrives, the machine can check whether one of the suspicious words shows up in it, and foresee its name in like manner. Such a framework would possibly be capable accurately to anticipate the name of inconspicuous messages [6, 7].

Correspondence in remote medium is dependably a channel and a great deal of research has been completed toward this path. This chapter centers around satellite correspondence with security issues that are settled utilizing machine learning approaches. In any case, there are no adequate investigations that emphasis on the qualities in remote system, particularly in low earth circle (LEO) satellite system. Topographically, an area on earth is described as an interwoven of concentrated land-employments [8–10]. Land-use is the reasonable and wise methodology of apportioning accessible land assets for various exercises, (for example, settlements, arable fields, pastures, and oversaw woods) inside an area. It is a method for using the land, including the portion, arranging, and the executives of its assets. The utilization of a specific fix of land and its physical character are connected. In any case, look into that builds up this connection is missing in spite of the expansion of geospatial data. This part further checks performances of GEO, LEO and MEO satellite system, and after that the nature of administration (QoS) measures under self-comparative traffic, for example, line length and cushion estimate are additionally talked about [11, 12].

Worldwide Positioning Systems (GPS) are utilized for acquiring the situation of vessels. Worldwide Positioning Systems, contrasted with the other framework are increasingly exact, trustable and valuable. Be that as it may, utilizing this framework isn't anchor against outside assaults consequently it can't be utilized on basic circumstances. In this examination, security of satellite frameworks has been stressed with arrangement which predicts the area of elements by handling input pictures of vessels, for example, radar pictures, real pictures or satellite pictures, to prepare framework on likeness metric has been advertised. Picture preparing world [13] as of late has been making a mind blowing progress on numerous troublesome issues by using profound learning strategies.

Ongoing advances in machine learning have prompted imaginative applications and administrations that utilization computational structures to reason about complex wonder. In the course of recent years, the security and machine-learning networks have created novel methods for developing ill-disposed examples and vindictive data sources made to deceive and subsequently degenerate the respectability of frameworks based on computationally learned models. The basic reasons for ill-disposed examples and the future countermeasures has been broke down [14] that may moderate them. This section will concentrate on the accompanying issues identified with security in Satellite correspondences in remote systems [15, 16].

Machine Learning is a field which is raised out of Artificial Intelligence(AI). Applying AI, we needed to manufacture better and keen machines. Be that as it may, aside from couple of simple errands, for example, finding the briefest way between two points, it isn't to program more mind boggling and continually developing difficulties. There was an acknowledgment that the best way to have the capacity to accomplish this undertaking was to give machine a chance to gain from itself. So machine learning was produced as another capacity for computers. Also, now machine learning is available in such huge numbers of sections of technology, that we don't understand it while utilizing it.

Machine learning (ML) is concerned about the structure and advancement of network security and strategies that enables systems to learn and train. The significant focal point of machine learning explore is to extricate data from information consequently, by computational and measurable techniques. It is subsequently firmly identified with information mining and insights. The intensity of neural networks originates from their portrayal ability. From one viewpoint, feed forward networks are demonstrated to offer the ability of general capacity guess. Then again, intermittent networks utilizing the sigmoidal initiation work are Turing proportionate and recreates a general Turing machine [15, 17]; Thus, repetitive networks can figure whatever work any advanced computer can register.

## 2 Cognitive Satellite Communication Issues Using Machine Learning

Future spacecraft communication subsystems will conceivably profit by programming characterized radios controlled by man-made reasoning calculations. A novel radio asset distribution calculation [18] utilizing multiobjective support learning and counterfeit neural system troupes ready to oversee accessible assets and clashing mission-based objectives. The vulnerability in the execution of thousands of conceivable radio parameter blends and the dynamic conduct of the radio channel after some time delivering a ceaseless multidimensional state-activity space requires a settled size memory nonstop state-activity mapping rather than the customary discrete mapping [16, 21]. What's more, activities should be decoupled from states so as to take into account internet learning, execution checking, and asset portion expectation. The proposed methodology use the creators' past research on obliging choices anticipated to have poor execution through "virtual condition investigation." The reproduction results demonstrate the execution for various communication mission profiles, and precision benchmarks are accommodated the future research reference. The proposed methodology establishes some portion of the center intellectual motor confirmation of-idea conveyed to the NASA John H. Glenn Research Center's SCaN Testbed radios on-board the International Space Station.

## 2.1 Satellite Communication Channel from Earth to GEO Orbit

Communication subsystems of space in future and related investigation missions can possibly profit by programming characterized radios (SDRs) controlled by machine learning calculations. In this paper, we propose a novel mixture radio asset allotment the board control calculation that incorporates multi-target support learning and profound fake neural networks. The goal is to effectively oversee communications framework assets by observing execution capacities with normal ward factors that bring about clashing objectives [19, 20]. The vulnerability in the execution of thousands of various conceivable blends of radio parameters makes the exchange off among investigation and abuse in fortification learning (RL) considerably more trying for future basic space-based missions. In this manner, the framework ought to invest as meager energy as conceivable on investigating activities, and at whatever point it investigates an activity, it ought to perform at adequate dimensions more often than not. The proposed methodology [22] empowers on-line learning by collaborations with the earth and confines poor asset designation execution through 'virtual condition investigation'. Enhancements in the multi-target execution can be accomplished by means of transmitter parameter adjustment on a parcel premise, with ineffectively anticipated execution immediately bringing about rejected choices. Recreations exhibited in this work considered the DVB-S2 standard versatile transmitter parameters and extra ones anticipated that would be available in future versatile radio frameworks. Execution results are given by investigation of the proposed crossover calculation while working over a satellite communication channel from Earth to GEO circle amid clear sky conditions. The proposed methodology comprises some portion of the center subjective motor confirmation of-idea to be conveyed to the NASA Glenn Research Center SCaN Testbed situated on-board the International Space Station.

## 2.2 Land Cover Prediction from Satellite Imagery Using Machine Learning

Different machine learning systems, for example, nearest neighbor algorithm, decision tree, support vector machine (SVM), random forest, naïve bayes classifier has been used [23] for arrive cover forecast from satellite symbolism. The information highlights are gathered from satellite picture utilizing normalized standardized distinction vegetation index (NDVI). The yield for six class arrangements is impenetrable, backwoods, plantation, homestead, grass and water [24, 25]. To adjust the information in each class engineered minority oversampling procedure (SMOTE) has been utilized. All the work has been completed utilizing python programming. The most astounding exactness is gotten utilizing k-NN.

## 2.3 *Performance Analysis of LEO Satellite Networks*

Based on communication process two Stochastic Petri Net (SPN) models are developed [26] to break down the execution of LEO Satellite Networks with one client and two clients individually. At that point, the impact of landing rate all things considered time delay is additionally investigated by fathoming the straight conditions from relating isomorphic Markov Chains of the SPN demonstrate under various parameters. The proposed methodology of demonstrating and execution assessment has extraordinary advantage to the structure and execution advancement of satellite networks [27].

Fault Prediction for Satellite Communication Equipment Based on Deep Neural Network-

With an objective of fault detection in satellite communication system, a forecast model dependent on deep learning is proposed [28]. Initially, the gear parameters are summed up, and after that two kinds of states covering ordinary and strange circumstances are resolved. After component learning, self-encoding network is utilized to get new highlights which can portray the profound element of the information. At that point the labeled information removed from checking gear are connected to prepare the expectation classifier which is the blend of profound conviction network and softmax classifier [29–32]. The profound conviction network is made out of restricted Boltzmann machine and additionally BP network. BP network is utilized for parameters change. At last, the impacts of blame expectation including the execution of model and normal forecast exactness are tried through simulation.

## 2.4 *An Adaptive Routing Based on an Improved ACO Technique in Leo Satellite Networks*

For adaptive routing, Ant colony optimization (ACO) has been proposed as a promising technique in communication networks. The calculation is by and large effectively connected to optimization issues in an assortment of fields. The first ACO has the disadvantages of stagnation conduct and moderate assembly. Proposition [33] sticles and enhances the variants of the first ACO so as to give better exhibitions. The enhanced directing calculation is mimicked in Iridium satellite group of stars. The outcomes demonstrate that the enhanced ACO not just accomplishes quick assembly in unique topology networks, yet additionally can keep away from networks clog and counterpoise the heap of the network.

## 2.5 *Rainfall Estimation Using Carrier to Noise of Satellite Communication*

In up coming machine learning approaches, data handling involves a vital job, and the learning control is utilized to find and take in information or properties from the data. Streamlining an execution paradigm utilizing model data and past experience gives a simple however dependable depiction about machine learning [1]. The quality or amount of the dataset will influence the learning and expectation execution. Machine learning likewise alluded to as Learning from Data, which accentuates the significance of data in numerous perplexing applications.

## 2.6 *Deep Learning for Amazon Satellite Image Analysis*

Machine learning strategies can be the used as the means to scant the world from losing miniature sized zones each second. As deforestation in the Amazon basin causes destroying impacts both on the biological community and nature, there is dire need to more readily comprehend and deal with its evolving scene. A Competition was recently organised to develop to breakmethodology down satellite pictures of the Amazon [32, 34, 35]. Successful strategies will have the capacity to identify unpretentious highlights in various picture scenes, giving us the significant information should have been ready to oversee deforestation and its outcomes all the more successfully [2].

## 2.7 *Satellite Super Resolution Images Using Deep Learning*

The deep learning neural network is an ongoing advancement that has turned into the subject of research in the PC vision and remote sensing disciplines. Super Resolution (SR) pictures can be acquired utilizing deep neural network techniques that accomplish a higher execution than all past customary strategies. Here, in this examination, the goal is to depict existing deep learning strategies for SR satellite pictures. Diverse satellite information are utilized to foresee the execution of every deep learning model. An a short review of most deep learning procedures and thinks about them to get a progressively powerful and productive model has been portrayed [36]. Another Adaptive Coding and Modulation (ACM) protocol [37] has been planned. The deep network course demonstrate other deep learning methodes this estimate is trustworthy in the recreation procedure for acquiring SR pictures and beats a few drawbacks found in customary reproduction calculations. The inadequate coding network technique stays significant, and with a few upgrades, further enhancement in results can be accomplished. Table 1 shows Description of Machine Learning techniques associated with other technology in network security.

**Table 1** Description of machine learning associated with other technology in network security

| Sl. No | Literature | Year | Network issue/challenge/security | Associated technology |
|---|---|---|---|---|
| 1 | P. McDaniel [14] | 2016 | Machine learning in adversarial environment | Machine learning |
| 2 | Y. Xin et al. [3] | 2018 | Machine learning and deep learning for cyber security | Cyber security |
| 3 | Q. Liu et al. [4] | 2018 | Security threats and defensive techniques of machine learning | Data science |
| 4 | M. Mozaffari et al. [17] | 2015 | Poisoning attack and defence using machine learning approach in health care | Health care |
| 5 | N. Islam et al. [16] | 2017 | Mobile phone security using Machine learning | Device security |
| 6 | Ahmad et al. [21] | 2018 | Support vector machine in random forest and extreme learning machine for IDS | Extreme learning and IDS |
| 7 | A. L. Buczak et al. [19] | 2016 | Data mining and ML for cyber security and IDS | Data mining |
| 8 | J. Wang et al. [20] | 2008 | State of the art about machine learning | Artificial intelligence |
| 9 | D. He et al. [24] | 2018 | Antenna selection for transmission in MIMO wiretap channels | Antenna transmission |
| 10 | X. Chen et al. [25] | 2018 | Deep learning and CNN for action recognition | DL and CNN |
| 11 | M. Ozay et al. [38] | 2016 | ML methods for attack detection in smart grid | Smart grid |
| 12 | N. Nissim et al. | 2017 | Unknown malicious Ms-Office document detection using active learning based on feature extraction | Image processing |
| 13 | X. P. Liu et al. | 2012 | Machine learning with kernel machine | |
| 14 | Y. Zheng et al. | 2017 | Airline passenger profiling based on fuzzy deep machine learning | Fuzzy logic deep learning |

## 3 Security and Prevention from Cyber Attacks in Satellite Communication

The development and advancement of cyber-attacks require strong and developing cyber security plans. As a developing innovation, the Internet of Things (IoT) acquires cyber-attacks and dangers from the IT condition in spite of the presence of a layered guarded security instrument. The augmentation of the computerized world to the physical condition of IoT brings inconspicuous attacks that require a novel lightweight and conveyed attack detection system because of their engineering and asset limitations. Compositionally, Fog computing based mobile stations can be utilized to offload security capacities from IoT and the cloud to moderate the asset restriction issues of IoT and versatility bottlenecks of the cloud. Traditional machine learning calculations have been widely utilized for intrusion detection, despite the fact that versatility, highlight designing endeavours, and precision have prevented their infiltration into the security advertise. These inadequacies could be alleviated utilizing the profound learning approach as it has been fruitful in huge information fields. Aside from disposing of the need to create includes physically, profound learning is strong against transforming attacks with high detection exactness. Diro et al. [39] proposed a LSTM arrange for circulated cyber-attack detection in mist to-things communication. Critical attacks have been investigated and dangers focusing on IoT gadgets were distinguished particularly attacks abusing vulnerabilities of remote correspondences. The directed investigations on two situations show the adequacy and productivity of more profound models over conventional machine learning models.

### *3.1 Non-reliable Data Source Identification Using Machine Learning Algorithm*

Recent advances in machine learning have prompted imaginative applications and administrations that utilization computational structures to reason about complex marvel. In the course of recent years, the security and machine-learning networks have created novel methods for developing ill-disposed examples—malicious data sources made to deceive and in this manner degenerate the trustworthiness of frameworks based on computationally learned models. The hidden reasons for antagonistic examples and the future countermeasures has been broke down [14] that may relieve them.

### 3.2 *Deep Learning and Machine Learning for Interruption in Network*

With the improvement of the Internet, cyber-attacks are changing quickly and the cyber security circumstance isn't idealistic. Overview report by Xin et al. [3] clarifies the key writing studies on machine learning (ML) and deep learning (DL) techniques for system enquiry of interruption identification and gives a concise instructional exercise portrayal of every ML/DL strategy. Distinctive security approaches were ordered and outlined dependent on their transient or warm connections. Since information are so essential in ML/DL strategies, it portrays a portion of the generally utilized system datasets utilized in ML/DL, talk about the difficulties of utilizing ML/DL for cyber security and give recommendations to inquire about bearings.

### 3.3 *Security Protected Procedures Using Machine Learning*

Machine learning is a standout amongst the most overall procedures in software engineering, and it has been generally connected in picture preparing, regular dialect handling, design acknowledgment, cyber security, and different fields. Notwithstanding fruitful utilizations of machine learning calculations in numerous situations, e.g., facial acknowledgment, malware location, programmed driving, and interruption discovery, these calculations and comparing preparing information are helpless against an assortment of security dangers, initiating a critical execution diminish. Consequently, it is indispensable to call for further consideration with respect to security dangers and comparing guarded procedures of machine learning, which persuades a complete review [4]. Up to this point, specialists from the scholarly community and industry have discovered numerous security dangers against an assortment of learning calculations, including credulous Bayes, strategic relapse, choice tree, bolster vector machine (SVM), rule part examination, bunching, and winning profound neural systems.

There are many implementations of machine learning approach that utilizes supervisory learning. In supervised learning, the framework attempts to gain from the past precedents that are given. (Then again, in unsupervised learning, the framework endeavors to discover the examples straightforwardly from the model given.) Speaking scientifically, regulated learning is the place you have both info factors (x) and yield variables (y) and can utilize a calculation to get the mapping capacity from the contribution to the yield. Regulated learning issues can be additionally partitioned into two sections, in meticulous characterization, and relapse.

A classification issue is the dilemma at which the yield variable is a classification or a gathering, for example, "dark" or "white" or "spam" and "no spam". Regression: A regression issue is the point at which the yield variable is a genuine esteem, for example, "Rupees" or "stature." Unsupervised Learning—In unsupervised learning, the calculations are left to themselves to find fascinating structures in the informa-

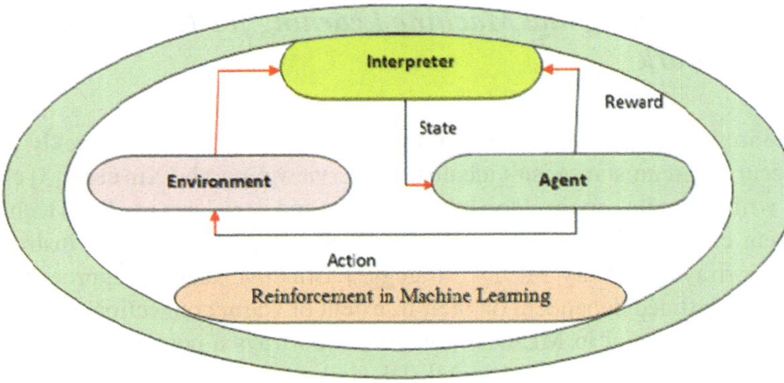


**Fig. 1** Reinforcement in machine learning

tion. Scientifically, unsupervised learning is the point at which you just have input information (X) and no relating yield factors. This is called unsupervised learning in light of the fact that not at all like directed learning above, there are no given right answers and the machine itself finds the appropriate responses. Unsupervised learning issues can be additionally separated into association and grouping issues. Association: An association rule learning issue is the place you need to find decides that depict substantial parts of your information [40–43], for example, "individuals that purchase X additionally will in general purchase Y". A clustering issue is the place you need to find the innate groupings in the information, for example, gathering clients by buying conduct.

## *3.4 Reinforcement Learning*

A computer program will communicate with a dynamic situation in which it must play out a specific objective, (for example, playing a diversion with a rival or driving a vehicle). The program is given criticism regarding prizes and disciplines as it explores its concern space. Utilizing this algorithm, the machine is prepared to settle on explicit choices. It works along these lines: the machine is presented to a situation where it consistently prepares itself utilizing experimentation technique (Fig. 1).

Machine Learning supposition is a field that meets factual, probabilistic, computer science and algorithmic angles emerging from learning drearily from information which can be utilized to assemble savvy applications. The preeminent inquiry when attempting to comprehend a field [44–47], for example, Machine Learning is the measure of maths important and the unpredictability of maths required to comprehend these frameworks. The response to this inquiry is multidimensional and relies upon the dimension and enthusiasm of the person. Here is the base dimension of science that is required for Machine Learning Engineers/Data Scientists. Machine learning approaches are basically used in mathematical fields such as linear algebra including matrix operations, projections, factorisation, symmetric matrix and orthogonalisation. In Probability and statistics it includes rules and axioms, bayes'theorem, random variables, variance, expectation, conditional and joint distributions. In calculus, differential and integral calculus and partial derivatives are implemented in machine learning approachs. Further Design of Algorithm and complex optimisations includes binary tree, hashing, heap and stack operations.

## 3.5 Extreme Learning Machine

It is obvious the learning speediness of feed forward neural networks is all in all far slower than required and it has been a noteworthy bottleneck in their applications for past decades. Two key purposes for might be: (1) the moderate gradient based learning calculations are broadly used to prepare neural networks, and (2) every one of the parameters of the networks are tuned ordinarily by utilizing such learning calculations. Table 2. Demonstrates Machine Learning and allied technology towards network security.

FFNN (Feed forward Neural Networks) are most widely utilized in numerous fields because of their capability such as (1) to estimated complex nonlinear mappings straightforwardly from the information tests; and (2) to give models to a substantial class of characteristic and counterfeit wonders that are hard to deal with utilizing traditional parametric methods. Then again, there need quicker learning calculations for neural networks. The conventional learning calculations are more often than not far slower than required. It isn't astonishing to see that it might take a few hours, a few days, and significantly more opportunity to prepare neural networks by utilizing customary techniques.

## 3.6 Malware Detection Using Machine Learning

In spite of the huge enhancement of digital security instruments and their ceaseless advancement, malware are still among the best dangers in the internet. Malware examination applies methods from a few distinct fields, for example, program investigation and network examination, for the investigation of pernicious examples to build up a more profound comprehension on a few viewpoints, including their conduct and how they advance after some time [48]. Inside the constant weapons contest between malware designers and experts, each development in security technology is normally speedily pursued by a relating avoidance. Some portion of the viability of novel cautious measures relies upon what properties they use on. For instance, a recognition rule dependent on the MD5 hash of a known malware can be effortlessly evaded by applying standard systems like jumbling, or further developed methodologies, for example, polymorphism or changeability. For a complete survey of these procedures. These techniques change the double of the malware, and hence its hash,

**Table 2** Depiction of machine learning and allied technology towards network security

| Sl. No | Literature | Year | Network issue/challenge/security | Associated technology |
|---|---|---|---|---|
| 1 | R. J. Mangialardo et al. [15] | 2015 | Static and dynamic malware analysis using machine learning | Network security |
| 2 | S. Earley et al. | 2015 | Analytics. Machine learning and IoT | Internet of things |
| 3 | S. Kalyani et al. | 2011 | Assessment and classification of power system security using multi-class SVM | Support vector machine |
| 4 | L. K. Shar et al. | 2015 | Web-app security using hybrid program analysis and machine learning | Web based application |
| 5 | A. Diro et al. [39] | 2018 | Fog to things communication and leveraging LSTM network | Fog computing |
| 6 | L. Han et al. | 2015 | Rule extraction from SVM using ensemble learning strategy | Ensemble learning strategy |
| 7 | H. Yan et al. | 2015 | Prototype based discriminative feature learning | Feature learning |
| 8 | S. Akcay et al. | 2018 | Deep CNN (convolution neural network) for object classification | Deep CNN |
| 9 | C. Yin et al. | 2017 | Deep learning for intrusion detection using recurrent neural network | recurrent neural network |
| 10 | R. Zhang et al. | 2012 | Extreme learning machine with adaptive growth of hidden nodes | Extreme learning |

yet leave its conduct unmodified. On the opposite side, creating identification decides that catch the semantics of a noxious example is considerably more hard to evade, in light of the fact that malware engineers ought to apply more mind boggling changes [6, 7]. A noteworthy objective of malware investigation is to catch extra properties to be utilized to enhance safety efforts and make avoidance as hard as would be prudent [38]. Machine learning is a characteristic decision to help such a procedure of information extraction. In fact, numerous works in writing have taken this bearing, with an assortment of methodologies, goals and results.

## 4 Conclusion

Malware investigation and categorization Systems utilize static and dynamic methods, related to machine learning calculations, to computerize the assignment of ID and grouping of malevolent codes. The two procedures have shortcomings that permit the utilization of analysis avoidance systems, hampering the ID of malwares. Mangialardo et al. [15] propose the unification of static and dynamic analysis, as a strategy for gathering information from malware that reductions the possibility of achievement for such avoidance strategies. From the information gathered in the analysis stage, we utilize the C5.0 and Random Forest machine learning calculations, actualized inside the FAMA structure, to play out the distinguishing proof and order of malwares into two classes and various classifications. The examinations and results demonstrated that the exactness of the bound together analysis accomplished a precision of 95.75% for the double arrangement issue and an exactness estimation of 93.02% for the different order issue. In all examinations, the brought together analysis created preferred outcomes over those acquired by static and dynamic breaks down detached.

## References


1. A. Gharanjik, M.R.B. Shankar, F. Zimmer, B. Ottersten, Centralized rainfall estimation using carrier to noise of satellite communication links. IEEE J. Sel. Areas Commun. **36**(5), 1065–1073 (2018). https://doi.org/10.1109/jsac.2018.2832798
2. L. Bragilevsky, I.V. Bajić, "Deep learning for Amazon satellite image analysis, in *IEEE Pacific Rim Conference on Communications, Computers and Signal Processing (PACRIM)*, (Victoria, BC, 2017) pp. 1–5. https://doi.org/10.1109/pacrim.2017.8121895
3. Y. Xin et al., Machine learning and deep learning methods for cybersecurity. IEEE Access **6**, 35365–35381 (2018). https://doi.org/10.1109/access.2018.2836950
4. Q. Liu, P. Li, W. Zhao, W. Cai, S. Yu, V.C.M. Leung, A survey on security threats and defensive techniques of machine learning: a data driven view. IEEE Access **6**, 12103–12117 (2018)
5. M. Rath, B. Pati, C.R. Panigrahi, J.L. Sarkar, QTM: A QoS task monitoring system for mobile ad hoc networks, in *Recent Findings in Intelligent Computing Techniques*, ed by P. Sa, S. Bakshi, I. Hatzilygeroudis, M. Sahoo. Advances in Intelligent Systems and Computing, vol 707 (Springer, Singapore, 2019)
6. M. Rath, B. Pati, B.K. Pattanayak, An overview on social networking: design, issues, emerging trends,and security, in *Social Network Analytics: Computational Research Methods and Techniques*, (Academic Press, Elsevier, 2018), pp. 21–47
7. M. Rath, J. Swain, B. Pati, B.K. Pattanayak,"Attacks and Control in MANET, in *Handbook of Research on Network Forensics and Analysis Techniques*. IGI Global, (2018), pp. 19–37
8. M. Rath, B. Pati, B.K. Pattanayak, Energy efficient MANET protocol using cross layer design for military applications. Defense Sci. J. 66(2) (DRDO Publication, 2016)
9. M. Rath, B. Pati, B.K. Pattanayak, Comparative analysis of AODV routing protocols based on network performance parameters in Mobile Adhoc Networks, in *Foundations and Frontiers in Computer, Communication and Electrical Engineering*. (CRC Press, Taylor & Francis, 2016), pp. 461–466. ISBN: 978-1-138-02877-7
10. M. Rath, C.R. Panigrahi, Prioritisation of security measures at the junction of MANET and IoT, in *Second International Conference on Information and Communication Technology*

*for Competitive Strategies*. (ACM Publication, New York, USA, 2016) http://www.acm.org/publications. ISBN: 978-1-4503-3962-9
11. M. Rath, B. Pati, B.K Pattanayak, Energy competent routing protocol design in MANET with real time application provision. Int. J. Bus. Data Comm. Network. IGI Global 11(1), 50–60 (2015)
12. M. Rath, Delay and power based network assessment of network layer protocols in MANET, in *2015 International Conference on Control, Instrumentation, Communication and Computational Technologies (IEEE ICCICCT)*. (Kumaracoil, India, 2015), pp. 682–686
13. M.M. Kiliç, Y.S. Akgül, Ship location estimation from radar and optic images using metric learning, in *2018 26th Signal Processing and Communications Applications Conference (SIU)*, Izmir (2018), pp. 1–4
14. P. McDaniel, N. Papernot, Z.B. Celik, Machine learning in adversarial settings. IEEE Secur. Priv. **14**(3), 68–72 (2016)
15. R.J. Mangialardo, J.C. Duarte, Integrating static and dynamic malware analysis using machine learning. IEEE Lat. Am. Trans. **13**(9), 3080–3087 (2015)
16. N. Islam, S. Das and Y. Chen, On-device mobile phone security exploits machine learning. IEEE Pervasive Comput. **16**(2), 92–96 (2017)
17. M. Mozaffari-Kermani, S. Sur-Kolay, A. Raghunathan, N.K. Jha, Systematic poisoning attacks on and defenses for machine learning in healthcare. IEEE J. Biomed. Health Inform. **19**(6), 1893–1905 (2015)
18. P.V.R. Ferreira et al., Multi objective reinforcement learning for cognitive satellite communications using deep neural network ensembles. IEEE J. Sel. Areas Commun., **36**(5). 1030–1041 (2018)
19. A.L. Buczak, E. Guven, A survey of data mining and machine learning methods for cyber security intrusion detection. IEEE Commun. Surv. Tutor. vol. **18**(2), 1153–1176 (2016)
20. J. Wang, Q. Tao, Machine learning: the state of the art. IEEE Intell. Syst. **23**(6), 49–55 (2008)
21. Ahmad, M. Basheri, M.J. Iqbal, A. Rahim, Performance comparison of support vector machine, random forest, and extreme learning machine for intrusion detection. IEEE Access **6**, 33789–33795 (2018). https://doi.org/10.1109/access.2018.2841987
22. P.V.R. Ferreira et al., Multi-objective reinforcement learning-based deep neural networks for cognitive space communications, in *2017 Cognitive Communications for Aerospace Applications Workshop (CCAA)*, Cleveland, OH (2017), pp. 1–8. https://doi.org/10.1109/ccaaw.2017.8001880
23. A. Panda, A. Singh, K. Kumar, A. Kumar, Uddeshya, A. Swetapadma, Land cover prediction from satellite imagery using machine learning techniques, in *Second International Conference on Inventive Communication and Computational Technologies (ICICCT)*, Coimbatore (2018), pp. 1403–1407. https://doi.org/10.1109/icicct.2018.8473241
24. D. He, C. Liu, T.Q.S. Quek, H. Wang, Transmit antenna selection in MIMO wiretap channels: a machine learning approach. IEEE Wirel. Commun. Lett. **7**(4), 634–637 (2018)
25. X. Chen, J. Weng, W. Lu, J. Xu, J. Weng, Deep manifold learning combined with convolutional neural networks for action recognition. IEEE Trans. Neural Netw. Learn. Syst. **29**(9), 3938–3952 (2018)
26. W. Zeng, Z. Hong, SPN-based performance analysis of LEO satellite networks with multiple users, in *International Conference on Machine Learning and Cybernetics*, Guilin ( 2011), pp. 1425–1429. https://doi.org/10.1109/icmlc.2011.6016850
27. M. Rath, B.K. Pattanayak, B. Pati, Energetic routing protocol design for real-time transmission in mobile ad hoc network, in *Computing and Network Sustainability*, Lecture Notes in Networks and Systems, vol 12. (Springer, Singapore, 2017)
28. T. Liu, K. Kang, H. Sun, Fault prediction for satellite communication equipment based on deep neural network, in *International Conference on Virtual Reality and Intelligent Systems (ICVRIS)*, Changsha (2018), pp. 176–178. https://doi.org/10.1109/icvris.2018.00050
29. M. Rath, B.K. Pattanayak, SCICS: a soft computing based intelligent communication system in VANET. Smart Secure Systems – IoT and Analytics Perspective. Communications in Computer and Information Science, vol 808. Springer (2018)

30. M. Rath, G.S. Oreku, Security issues in mobile devices and mobile adhoc networks,in *Mobile Technologies and Socio-Economic Development in Emerging Nations* (IGI Global, 2018), p. 80, ISBN 152254030X. DOI-https://doi.org/10.4018/978-1-5225-4029-8.ch009
31. M. Rath, An analytical study of security and challenging issues in social networking as an emerging connected technology (20 Apr 2018). In *Proceedings of 3rd International Conference on Internet of Things and Connected Technologies (ICIoTCT)*, 2018 held at Malaviya National Institute of Technology, Jaipur (India) on 26–27 Mar 2018
32. M. Rath, J. Swain, IoT security: a challenge in wireless technology. Int. J. Emerg. Technol. Adv. Eng. **8**(4), 43–46 (2018), April ISSN: 2250 – 2459 (Online)
33. Z. Gao, Q. Guo, P. Wang, An adaptive routing based on an improved ant colony optimization in leo satellite networks, in *2007 International Conference on Machine Learning and Cybernetics*, Hong Kong (2007), pp. 1041–1044. https://doi.org/10.1109/icmlc.2007.4370296
34. M. Burmester, B. de Medeiros, On the security of route discovery in MANETs. IEEE Trans. Mob. Comput. **8**(9), 1180–1188 (2009)
35. M. Carvalho, security in mobile ad hoc networks. IEEE Secur. Priv. **6**(2), 72–75 (2008)
36. H.M. Keshk, X. Yin, Satellite super-resolution images depending on deep learning methods: a comparative study, in *2017 IEEE International Conference on Signal Processing, Communications and Computing (ICSPCC)*, Xiamen (2017), pp. 1–7. https://doi.org/10.1109/icspcc.2017.8242625
37. A. Tsakmalis, S. Chatzinotas, B. Ottersten, Automatic modulation classification for adaptive power control in cognitive satellite communications, in *7th Advanced Satellite Multimedia Systems Conference and the 13th Signal Processing for Space Communications Workshop (ASMS/SPSC)*, Livorno (2014), pp. 234–240
38. M. Ozay, I. Esnaola, F.T. Yarman Vural, S.R. Kulkarni, H.V. Poor, Machine learning methods for attack detection in the smart grid. IEEE Trans. Neural Netw. Learn. Syst. **27**(8), 1773–1786 (2016)
39. A. Diro, N. Chilamkurti, Leveraging LSTM networks for attack detection in fog-to-things communications. IEEE Commun. Mag. **56**(9), 24–130 (2018)
40. B. Rong, H. Chen, Y. Qian, K. Lu, R.Q. Hu, S. Guizani, A pyramidal security model for large-scale group-oriented computing in mobile ad hoc networks: the key management study. IEEE Trans. Veh. Technol. **58**(1), 398–408 (2009)
41. N. Saxena, G. Tsudik, J.H. Yi, Efficient node admission and certificateless secure communication in short-lived MANETs. IEEE Trans. Parallel Distrib. Syst. **20**(2), 158–170 (2009)
42. Y. Wang, F.R. Yu, H. Tang, M. Huang, A mean field game theoretic approach for security enhancements in mobile ad hoc networks. IEEE Trans. Wirel. Commun. **13**(3), 1616–1627 (2014)
43. U. Ghosh, R. Datta, A secure addressing scheme for large-scale managed MANETs. IEEE Trans. Netw. Serv. Manage. **12**(3), 483–495 (2015)
44. Z. Wei, H. Tang, F.R. Yu, M. Wang, P. Mason, Security enhancements for mobile ad hoc networks with trust management using uncertain reasoning. IEEE Trans. Veh. Technol. **63**(9), 4647–4658 (2014)
45. S. Surendran, S. Prakash, An ACO look-ahead approach to QOS enabled fault- tolerant routing in MANETs. China Commun. **12**(8), 93–110 (2015)
46. D.Q. Nguyen, M. Toulgoat, L. Lamont, Impact of trust-based security association and mobility on the delay metric in MANET. J. Commun. Netw. **18**(1), 105–111 (2016)
47. S.K. Dhurandher, M.S. Obaidat, K. Verma, P. Gupta, P. Dhurandher, Faces: friend-based ad hoc routing using challenges to establish security in MANETs systems. IEEE Syst. J. **5**(2), 176–188 (2011)
48. J. Chang, P. Tsou, I. Woungang, H. Chao, C. Lai, Defending against collaborative attacks by malicious nodes in MANETs: a cooperative bait detection approach. IEEE Syst. J. **9**(1), 65–75 (2015)
49. M. Rath, B.K. Pattanayak, Security protocol with ids framework using mobile agent in robotic MANET. Int. J. Inf. Secur. Priv. **13**(1), 46–58 (2019). https://doi.org/10.4018/ijisp.2019010104

50. M. Rath, B. Pati, B. Pattanayak, Manifold surveillance issues in wireless network and the secured protocol. Int. J. Inf. Secur. Priv.(IJISP) **13**(3), Article 3 (2019)
51. M. Rath, B. Pattanayak, Technological improvement in modern health care applications using Internet of Things (IoT) and proposal of novel health care approach. Int. J. Hum. Rights Healthc., ISSN: 2056-4902. (2018). https://doi.org/10.1108/ijhrh-01-2018-0007
52. M. Rath, Big data and iot-allied challenges associated with healthcare applications in smart and automated systems. Int. J. Strat. Inf. Technol. Appl. (IJSITA) **9**(2) (2018). DOI: https://doi.org/10.4018/ijsita.2018040l0
53. M. Rath, B. Pati (2017) Load balanced routing scheme for MANETs with power and delay optimisation, Int. J. Commun. Netw. Distrib. Syst. (IJCNDS) **19**. Inderscience Publishers
54. M. Rath, Resource provision and QoS support with added security for client side applications in cloud computing. Int. J. Inf. Technol. **9**(3), pp 1–8 (2017)
55. M. Rath, B.K. Pattanayak, Monitoring of QoS in MANET based real time applications, in *Information and communication technology for intelligent systems,* vol. 2, ICTIS, ed. by S. Satapathy, A. Joshi. Smart Innovation, Systems and Technologies, vol 84, pp. 579–586, Springer (2018)
56. M. Rath, B. Pati and B.K. Pattanayak, Cross layer based QoS platform for multimedia transmission in MANET, in *11th International Conference on Intelligent Systems and Control (ISCO)*, Coimbatore (2017), pp. 402–407
57. M. Rath, B. Pattanayak "MAQ: a mobile agent based QoS platform for MANETs. Int. J. Bus. Data Commun. Netw. **13**(1), 1–8 (2017). IGI Global
58. M. Rath, M.R. Panda, MAQ system development in mobile ad-hoc networks using mobile agents, in *IEEE 2nd International Conference on Contemporary Computing and Informatics* (IC3I), Noida (2017), pp. 794–798
59. S. Chaturvedi, V. Mishra, N. Mishra, Sentiment analysis using machine learning for business intelligence, in *IEEE International Conference on Power, Control, Signals and Instrumentation Engineering (ICPCSI)*, Chennai (2017), pp. 2162–2166
60. C. Feng, S. Wu, N. Liu, A user-centric machine learning framework for cyber security operations center, in *IEEE International Conference on Intelligence and Security Informatics (ISI)*, Beijing (2017), pp. 173–175

# Machine Learning Techniques for IoT Intrusions Detection in Aerospace Cyber-Physical Systems

**Yassine Maleh**

**Abstract** Aeronautical systems are no longer traditional masterpieces of autonomous mechanical engineering. Today, they are characterized by many intelligent technologies that include sensors, wireless standards and data analysis tools. Known as Aerospace Cyber-physical Systems (CPS), these CPSes are undergoing a massive transformation to increase the safety, efficiency and reliability of their operations. The physical system has created the Internet of Things IoT by integrating sensors, controllers and actuators. Nevertheless, the cyberspace of these aerospace CPSes offers many opportunities for malicious actors who threaten the security and privacy of vehicles/aircraft and their applications. Unprotected or poorly protected systems can easily be exploited for malicious purposes. Indeed, aerospace CPSes are always under threat from an increasing number of cyber-attacks through sensory or wireless channels, hardware, software or actuators. Recently, due to the significant advances and impressive results of machine learning techniques in the fields of image recognition, natural language processing and speech recognition for various long-standing artificial intelligence tasks, there has been a great interest in applying them to intrusion detection in the field of cybersecurity. In this chapter, we present different machine learning techniques for IoT intrusion detection in aerospace cyber-physical systems. The application of machine learning for cybersecurity in IoT requires the availability of substantial data on IoT attacks, but the lack of data on IoT attacks is a significant problem. In our study, the Cooja IoT simulator was used to generate high fidelity attack data in IoT 6LoWPAN networks. The efficient network architecture for all machine models is chosen based on comparing the performance of various network topologies and network scenarios. The experimental results show that Machine learning models for intrusion detection give better results by more than 99% in terms of accuracy, efficiency and detection rate. Also, it requires a low energy consumption overhead and memory, which proves that the proposed models can be used in constrained environments such as IoT sensors.

---

Y. Maleh (✉)
The National Port Agency, Casablanca, Morocco
e-mail: y.maleh@uhp.ac.ma

Faculty of Science and Technology, Hassan 1st University, Settat, Morocco



A. E. Hassanien et al. (eds.), *Machine Learning and Data Mining in Aerospace Technology*, Studies in Computational Intelligence 836,
https://doi.org/10.1007/978-3-030-20212-5_11

**Keywords** Cybersecurity · Internet of Things · Machine learning · Aerospace Cyber-Physical system · 6LoWPAN · Routing attacks · Intrusion detection

## 1 Introduction

In 2017, the world experienced one of the worst waves of cybersecurity attacks. From the data breach of the Equifax credit agency that affects more than 143 million consumers to the tyranny of the WannaCry ransom program that paralyzed several English hospitals and factories such as Renault in Sandouville or Slovenia, etc. However, the most alarming of them were for national defenses, such as the targeted attack by suspected Russian hackers on the US Army and NATO in October and the spyware campaign against the military and security organizations of the Indian and Pakistani governments.

Cybersecurity attacks are becoming more frequent as cyber attackers exploit system vulnerabilities for a financial gain [1]. Nation-state actors employ the most skilled attackers, capable of launching targeted and coordinated attacks. Sony, PumpUp and Saks, Lord & Taylor are recent examples of targeted attacks. The time between a security breach and detection is measured in days. Cyber attackers are aware of existing security controls and are continually improving their attack techniques [2]. To extend the range of attacks, cyber attackers have a wide range of tools at their disposal to bypass traditional security mechanisms. Malicious infection control frameworks, Zero-day exploits and rootkits can easily be purchased on an underground market. Attackers can also buy personal information and compromised domains to launch additional attacks [3].

As cyber attacks have become increasingly sophisticated, with allegations that one country is targeting another country for geopolitical purposes, the rate of investment in cybersecurity in the aerospace and defense market has also increased [4]. In these times of paranoia, governments and organizations are investing more than ever in the cybersecurity of defense and aerospace products and services. According to Netscribes' research, the contribution of the global defense and aerospace sector is expected to reach $24.37 billion by 2022. The overall size of the cybersecurity market is also expected to grow strongly over the same period, reaching $125 billion by 2025.

Nowadays, machine learning is one of the most popular topics to detect cyber-attacks on Internet if Things IoT. Because in-depth knowledge-based techniques can offer a robust system for sophisticated attacks. On the contrary, the biggest problem in IoT's security research is the lack of public and updated datasets.

Traditional machine learning techniques, such as Bayesian Belief Networks (BBN) [5–9] have been applied for cybersecurity. However, the generated large-scale data in IoT requires an efficient machine learning-based method, which can be adapted to the IoT specifications.

In this paper, we generated data by real-time simulations due to a lack of availability of public data sets for IoT attacks, and also existing datasets such as KDDCup 99

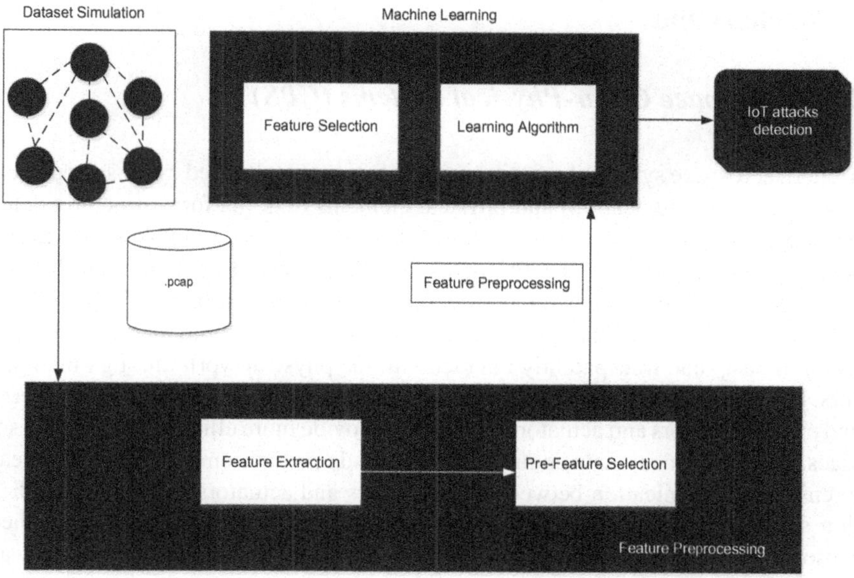


**Fig. 1** The proposed methodology

are too old. Simulation generated raw packet capture files are first converted to CSV (Comma Separated Values) files for processing and are then input into the feature preprocessing module of our system. We identified 28 characteristics as an initial set of features. Then, the normalization of characteristics is applied to all data sets to reduce the adverse effects of marginal values. As a result of this analysis, some features are abandoned in the pre-functional selection process. We have reduced the number of features to 16 main features. The approach used is described in Fig. 1.

The energy consumption and calculation capacities of its IoT devices are the most critical constraints of this type of network. Due to these constraints, the security solutions designed for IoT should be both lightweight and efficient, which would reduce the computational load on the devices as much as possible [10]. The objective of this chapter is to propose a lightweight solution that imposes a minimal load on the IoT network. The overall goals are:

- To generate a new IoT routing attacks dataset.
- To build different machine learning algorithms and train them by produced datasets.
- To evaluate the different ML models for intrusion detection in IoT.

This chapter presents the research background in the next section: the related work in Sect. 2 and the detailed description of the proposed intrusion detection methodology in Sect. 3. Section 4 describes the experiments using the proposed models. Section 5 presents conclusions and future research directions.

# 2 Background

## *2.1 Aerospace Cyber-Physical Systems (CPS)*

Modern aerospace systems have a strong link between embedded cyber systems (e.g. processing, communication) and physical elements (e.g. platform structure, detection, activation and environment). Researchers have begun to explore and exploit "cyber-physical systems" or CPS, defined as "technical systems that are built on and depend on the synergy of physical and computer components" [11]. These CPSs consist of interconnected systems of heterogeneous components can operate autonomously and transparently interact with the physical world through their sensors. For example, a commercial aircraft or a driverless car has thousands of internal and external sensors and actuators on board to provide more efficient and reliable services. Similarly, many new communication standards have emerged in recent years to ensure communication between these sensors and actuators for various application scenarios. Manufacturers can now collect huge amounts of data using these sensors to perform real-time operations and accurately identify hardware, software and communication failures.

Aerospace CPS contain critical data, conduct research and collaboration activities, and improve quality of life. CPSs are intelligent systems that provide an environment for the cooperation of computer components and things that are well known for their physical activities. The CPS Aerospace is a kind of bridge that brings the cybernetic and physical domains together and assumes an indispensable responsibility in many areas as clearly as possibly illustrated in Fig. 2.

The first CPS applications were based on smartphone devices to deploy applications. As a result, personal assistance applications have developed, particularly those focusing on medical assistance. The vision of "connected health" has grown in recent years, thanks in particular to the development of related technologies such as wireless networks or sensors. This has led to the development of Personal Health Devices (PHD), which aim to collect and share information on a local network or the Internet [12].

CPSs are cooperating systems, with decentralized control, resulting from the fusion between the real and virtual worlds, with autonomous and context-dependent behaviors, which can be constituted as systems with other CPSs and lead to extensive collaboration with humans [13]. For this purpose, CPS embedded software uses sensors and actuators, connects and with human operators by communicating via interfaces and can store and process information from sensors or the network [14]. According to Shi et al. [15], here are the characteristics attached to a CPS:

- High level of physical integration/cyber;
- Processing capabilities in each physical component, because processing and communication resources are generally limited;
- Highly connected, via networks with or without wires, Bluetooth, GSM, GPS, etc.;

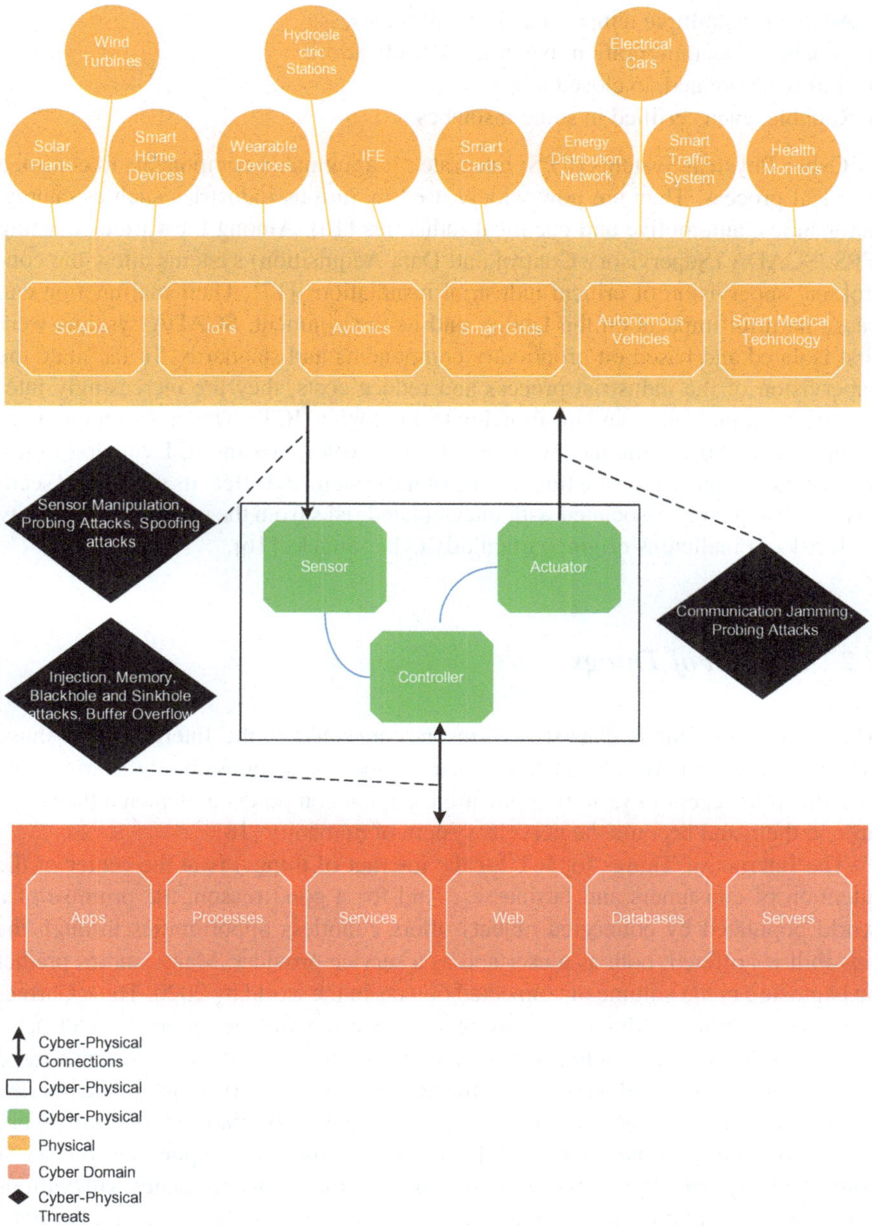


**Fig. 2** Aerospace cyber-physical system architecture

- Adapted to multiple temporal and spatial scales;
- Capable of reconfiguration/dynamic reorganization;
- Highly automated, in closed loops;
- Reliable, even certified in some instances.

Cyber-Physical Systems (CPS) integrate programmable components to control a physical process. They are now widely used in various industries such as energy, aeronautics, automotive and chemical industries [16]. Among the various existing CPS, SCADA (Supervisory Control and Data Acquisition) systems allow the control and supervision of critical industrial installations [17]. Their malfunction can cause harmful impacts on the facility and its environment. SCADA systems were first isolated and based on proprietary components and standards. To facilitate the supervision of the industrial process and reduce costs, they are increasingly integrating communication and information technologies (ICT). This makes them more complex and exposes them to cyber attacks that exploit existing ICT vulnerabilities. These attacks can change the functioning of the system and affect its safety, and security is subsequently associated with uncalculated risks from the system, and security with risks of malicious origin, particularly cyber-attacks [16].

## 2.2 *Internet of Things*

The Internet of Things consists of sensors connected to the Internet that behave similarly to the Internet by making open ad hoc connections, freely sharing data and allowing access to various applications so that computers understand the world around them and become the nervous system of humanity [18].

The Internet of Things (or IoT for the internet of things) is at the center of the attention of consumers and businesses. And for a good reason, the promise of a world populated by connected objects offers countless opportunities through the possibilities offered, both as a user and as a service provider. Many studies predict an explosion in the volume of connected objects in the world by 2020. Thus, Gartner forecasts 26 billion. Although a strong vigilance remains necessary to read these figures, as the perimeter definitions vary so much, they nevertheless confirm a trend towards the massive deployment of connected objects. The very notion of the Internet of Things, which is subject to interpretation, deserves to be clarified. For this report, a broad definition of the Internet of Things will be used, corresponding to a set of connected physical objects that communicate via multiple technologies with various data processing platforms, in connection with the waves of the cloud and big data. Data and its uses are at the heart of the Internet of Things. These, extracted from the various terminals and sensors, make it possible to inform users in real time of the evolution of their environment. Beyond the simple provision of information, the aggregation of the multiplicity of this data from heterogeneous sources makes it possible to quantify the connected environment to identify trends, enrich uses or consider new ones. The user—individual or company—can act in real time on his environment—manually or automatically—to optimize processes (for example, optimization of road flows or supply chains in real time).

The applications of the Internet of Things result in many concrete uses—new or improved—that have a significant impact on the daily lives of individuals, companies, and communities. The potential benefits expected to facilitate its adoption by this diversity of users. Several sectors, or growth markets, stand out, in particular:

- The so-called "intelligent" territories are at the heart of local authorities' projects and should make it possible to optimize the management of communicating infrastructures (transport, energy, water, etc.) to provide a better service to citizens and respect sustainable development objectives within the territories.
- Thanks to the Internet of Things, housing and workplaces are becoming more comfortable, easier to manage and less expensive to use. The connected building, including the connected house, offers in particular possibilities for controlling energy consumption, integrating security and comfort systems and increasing comfort;
- The industry of the future (the use of the Internet of Things to serve the means of production) is gradually developing. The first step is the transfer of information. Feedback and remote control are more complex phases to implement in some areas of activity;
- The connected vehicle, for which first applications have already been developed, has also taken the first step in reading the information thanks to the integration of long-standing on-board electronics. The actors of the automotive industry are now seeking to develop new business models to take advantage of these new opportunities while issues related to responsibilities are emerging;
- Connected health, including the "wellness" segment, is one of the applications to which the general public is most aware, mainly thanks to wearables. The aspects related to the protection of personal data focus attention, because of the collection of unusually intimate—or even health—and new personal information by private actors and the stakes involved in their exploitation, particularly by certain services. The technological contributions on the organization of care and the degree of involvement of health professionals is also a subject of attention. The changes made possible by technological developments that are often faster than social and regulatory developments make this sector more difficult to understand and more complex.

To invest this new field of Internet of Things, protocols must be adapted to new constraints; security must be reinforced because the objects have an effect in the real world and a malfunction can lead to serious consequences. As regards architecture, they must be the most generic possible to allow interconnection and they must not be linked to a particular purpose.

The 6LoWPAN protocol, for IPv6 Low power Wireless Persona Area Network, is an adaptation of the IPv4 and IPv6 protocols for communications involving connected objects [19]. It was developed by the Internet Engineering Task Force (IETF) to be "lighter" than standard IP protocols. Also running on a mesh network model, it fully supports UDP and TCP. Mulligan states that the packet headers are very light (2–11 bytes) and can allow communication between $2^{64}$ nodes [20]. Also, most of the work on IDS in connected objects has focused on this protocol (Zarpelão et al. 2017). Like the ZigBee, the 6LoWPAN operates in the 2.4 GHz frequency band, making its integration easier due to current equipment. Thus, the heterogeneity of connected objects within homes makes it much more difficult to propose generic solutions for securing connected objects in this context.

## *2.3 Security Overview in IoT*

The security of IoT systems can be exceptionally complex due to a large number of components, a potentially large attack surface and interactions between different parts of the system. Threat modeling is an excellent starting point for understanding the risks associated with IoT systems and how these risks can be mitigated. IoT security is important and routing attacks are a widespread threat to IoT [21].

RPL attacks can be classified into three categories according to the vulnerability they aim to exploit (resources, topology and traffic.) These resource-based attacks are DoS-based and aim to deplete energy and overload memory. Topological attacks are intended to interfere with the normal network process. This can lead to the failure of one or more network nodes. Also, these attacks threaten the original topology of the network. Traffic-based attacking nodes aim to join the network like a normal node. Then, these attackers use network traffic information to conduct the attack [22].

Routing attacks are most common against Routing Protocol for Low-Power and Lossy Networks (RPL). Among the most significant routing attacks are hello-flood, wormhole and Sinkhole attacks. Figure 3 illustrates the location of routing attacks in IoT for cyber physical environment.

### 2.3.1 Attacks and Threats in IoT

With the development of IoT, more devices are becoming connected to the internet. Every day, these devices are becoming target for several attacks [23]. To address the security challenges in IoT, the authors need to analyze the security problems in IoT based on four-layer architecture. There are different types of attacks on the IoT. These attacks can be active attacks in which an attacker attempts to make changes to data on the target or data in route to the target, or there can be passive attacks in which an attacker attempt to obtain or make use of information. The attacker can perform various attacks like network jamming, message sniffing, device compromising etc.

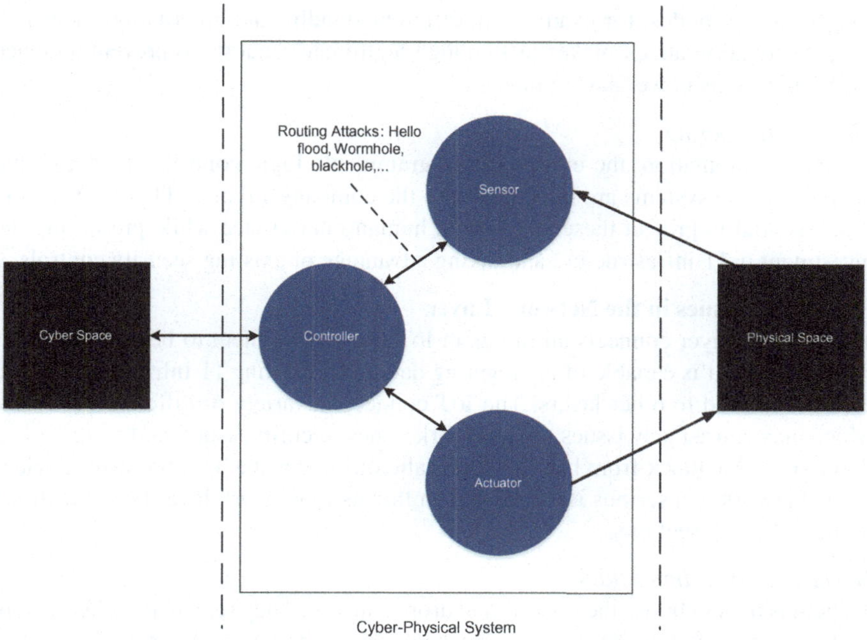


**Fig. 3** Routing attacks in IoT-CPS

**A. Security Issues in the Physical Layer**

There are many security issues affecting the physical layer of the IoT system. There is a great need for new technologies to protect energy resources and physical security mechanisms. The devices must be protected against physical attacks. They must also be able to save and optimize energy and be able to rely on battery power in the event of a power failure or interruption of the city's grid. The batteries must be charged long enough and recharged quickly so that the device can continue to operate [24]. Common issues in Physical layer been identified in the following sections.

*Physical Damage*

An example scenario in this type of attack is physical devices such as sensors, nodes and actuators that are physically damaged by malicious entities. This could cause the sensor, nodes and actuators to lose its expected functionality and become vulnerable to other risks.

*Environmental Attacks*

In this kind of attack, physical devices such as sensors, nodes and actuators are physically damaged by malicious entities. The sensor, nodes and actuators could thus lose the expected functionality and become vulnerable to other risks.

*Loss of Power*

Devices that lack energy cannot operate normally, resulting in a denial of service. For example, a common strategy for saving energy is to switch appliances to various energy-saving modes, for example, in different standby and hibernation modes. A sleep deprivation attack makes just enough legitimate requests to prevent a device from entering its power-saving mode.

*Physical Tampering*
In factory automation, the embedded programmable logic controllers (PLCs) that operate robotic systems are integrated into the company's typical IT infrastructure. It is essential to protect these PLCs from human interference while preserving the investment in IT infrastructure and taking advantage of existing security controls.

**B. Security Issues in the Network Layer**
The network layer connects all things in IoT and allows them to be aware of their surroundings. It is capable of aggregating data from existing IT infrastructures and then transmitted to other layers. The IoT connects a variety of different networks, which may cause many issues with network issues, security issues, and communication issues. An attack from hackers and malicious nodes that compromises devices in the network is a serious issue. Common threats to network layer been identified in the following sections.

*Selective Forwarding Attack*
Malicious nodes choose the packets and drop them out. They selectively filter certain packets and allow the rest. Dropped packets may carry the necessary sensitive data for further processing.

*HELLO Flood Attacks*
In HELLO flood attack, every node will introduce itself with HELLO messages to all the neighbors that are reachable at its frequency level. A malicious node will cover a wide frequency area, and hence it becomes a neighbor to all the nodes in the network. Subsequently, this malicious node will also broadcast a HELLO message to all its neighbors, affecting the availability.

*Sinkhole Attack*
In this attack, the malicious node advertises itself as the best path to be chosen as a preferred parent by its neighbors, and thus to route traffic through it. As it is, this attack does not appear to be harmful (passive attack). However, it becomes harmful (active attack) if combined with other attacks [25].

*Blackhole Attack*
An intruder triggers a black-hole attack by dropping all data packets routed through it. This attack can be considered as a DoS Attack. Indeed, the blackhole attack is more dangerous if combined with Sinkhole attacks since the attacker is in a position where massive traffic is routed through it. This attack increases the number of exchanged DIO messages which leads to instability of the network; data packets delay and thus resources exhausting [26].

*Selective-Forwarding Attack*
In selective redirection attacks, a malicious node can either actively filter RPL control messages or drop data packets and transfer only control message traffic. The first

attack negatively affects the construction of the topology and network functions, which disrupts routing. While the second attack leads to a DoS attack because no data will be transmitted to the destination nodes. These attacks are also known as grey hole attacks, which are a special case of blackhole attacks [25].

*Wormhole Attack*
To trigger a wormhole attack, two or multiple attackers have to connect via wired or wireless links called tunnels. Wormhole attack permits an attacker to replay the network traffic in the other ends of the tunnels. In the case of RPL, some attackers can be outside the 6LoWPAN, and thus can bypass the 6LBR. Also, if control messages are replayed to another part of the network, nodes that are distant see each other as if they are neighbors which leads to distorts routing paths and create un-optimized paths (Mayzaud et al. 2016).

*DoS Attack*
DoS attack aims to make nodes and/or the network unavailable. These attacks can be triggered against any layer of the IoT architecture. These attacks are simple to implement and very common because they have devastating consequences on the network.

*Storage Attacks*
Vast portions of data containing dynamic information of the user will need to be stored on storage devices, this one can be attacked and the data may be compromised or changed. The repetition of the data coupled with the access of data to different types of people results in the increased surface area for the attacks.

**C. Security Issues in the Perception Layer**
The security threats in the Perception layer are at the node level. Because the nodes are composed of sensors, they are prime targets for hackers who want to use them to replace the device's software with their software. In the perception layer, the majority of the threats comes from the outside entities, mostly concerning sensors and other data gathering utilities. Common threats in Perception Layer been identified in the following sections.

*Eavesdropping*
In wireless communication, the communication between devices is wireless and through the Internet, this makes them vulnerable to eavesdropping attacks.

An adversary can perform an attack scenario, for example, a sensor in the smart home that is compromised can send thrust notification to users and collect private information from the users.

*Sniffing Attacks*
To acquire information from the device, an attacker put malicious sensors or sniffers close to the normal sensors of the IoT devices. For example, as human-to-human and human-to-device interactions occur over shared physical networks, services and social spaces, it is also possible to detect smaller amounts of physical drag from these interactions with a higher degree of sensitivity and accuracy.

*Noise in Data*
As the data transmission over wireless networks covering vast distances, it is probable that the data may contain noise i.e., false information, missing information. Falsification of data can be dangerous in such scenarios when a lot is dependent on the reliable transmission of data.

**D. Security Issues in the Application Layer**
Due to security issues in the application layer, applications can be easily stopped and compromised. As a result, applications are not able to run the services for which they are programmed or even execute authenticated services incorrectly. In this layer, malicious attacks can cause bugs in the application program code that cause the application to malfunction. This is a very critical concern given the number of devices classified as entities at the application level. Threats common to the application layer have been identified in the following sections.

*Data Authentication*
Data can be collected from any device at any time. They can be falsified by intruders. It must be ensured that the perceived data comes only from intended or legitimate users. Also, it is mandatory to check that the data have not been modified during transit. Data authentication could ensure integrity and originality.

*Malicious Code Attacks*
An example of a scenario in this kind of attack could be a malicious worm spreading to embedded Internet attack devices running a particular operating system for Linux, for example [27]. Such a worm could be able to attack a range of small Internet-compatible devices, such as home routers, set-top boxes and security cameras. The worm would use a software vulnerability known to spread. Such code attacks could enter a car's Wi-Fi, take control of the steering wheel and cause the car to crash, injuring both the driver and the car.

*Tampering with Node-Based Applications*
Hackers exploit application vulnerabilities on device nodes and install malicious rootkits. The security design of the devices must be tamper-proof or at least tamper-proof. The protection of specific parts of a device may not be sufficient. Some threats can manipulate the local environment to cause the device to malfunction and cause the environment to heat or freeze. An altered temperature sensor would only display a fixed temperature value, while the altered camera in the smart house would transmit outdated images.

## 2.4 Machine Learning Techniques

Machine learning techniques are based on the establishment of an explicit or implicit model to categorize classification problems in the target system. A unique feature of these approaches is the need to provide strong data to form the behavioral model.

Depending on the organization of these data, we can classify them into three main categories:

- Supervised learning: Training data includes both input characteristics and output decision,
- Semi-supervised learning: The training data only contains the characteristics of the problem to be solved,
- Unsupervised learning: No training data is provided as input.

In many cases, the applicability of automatic learning principles coincides with that of statistical techniques; it focuses on building a model that improves its performance based on previous results. Therefore, a learning algorithm can modify its execution strategy based on new information about the problem to be solved. Although this characteristic may make it desirable to use such schemes for all situations, the major disadvantages are their resource-intensive nature during the learning phase and the sometimes-high error rates, as well as the non-explicit nature of the alarms raised by these models. Other phenomena can impact algorithms by automatic learning. Some algorithms such as decision trees and SVMs are often subject to the phenomenon of overlearning. Thus, by evaluating the performance indicators on the training data, we find a largely optimistic estimate of the classifier's performance. Below are the most commonly used models in the field of anomaly detection and their main advantages and disadvantages.

### 2.4.1 Bayesian Models

We distinguish two categories of Bayesian models: The simple or naive Bayesian and Bayesian networks. The first method is based on Bayesian inferences that allow the probability of an event to be deduced from those of other events already evaluated. Thus, they reduce the high-density estimate to a one-dimensional density estimate of the nucleus, using the assumption that the input characteristics are independent. While the second approach is a model that encodes probabilistic relationships between variables of interest. This technique is generally used for the detection of attacks in combination with statistical patterns, a procedure that provides several advantages [10], including the ability to code and predict interdependencies between variables. As well as the ability to integrate both knowledge and past data. However, as pointed out in [28], a serious disadvantage of using Bayesian networks is that their results are similar to those derived from threshold-based systems, while the computational effort is considerably higher. Although the use of Bayesian networks has proved effective in some situations, the results obtained are highly dependent on assumptions about the behavior of the target system and therefore a deviation in these assumptions Leads to detection errors, attributable to the model considered.

### 2.4.2 Decision Tree Models

A decision tree is a tree structure with leaves representing classifications and branches representing conjunctions of characteristics that lead to solving classification problems. A copy is labeled (classified) by testing its characteristic value (attribute) against the nodes of the decision tree. The most common methods for automatically building decision trees are ID3 [29] and C4.5 [30] algorithms. Both algorithms construct decision trees from a set of training data using the concept of entropy. When building the decision tree, at each node of the tree, C4.5 chooses the data attribute that most effectively divides its set of examples into subsets. The splitting criterion is the gain of normalized information (entropy difference). The attribute with the highest standardized information gain is chosen to make the decision. The C4.5 algorithm then recurs on the smaller subassemblies until all training examples have been classified. The advantages of decision trees are the intuitive expression of knowledge, high classification accuracy and ease of implementation. The main disadvantage is that for data including categorical variables with a different number of levels, the information gain values are biased in favor of characteristics with more levels. The decision tree is constructed by maximizing the gain of information at each fraction of a variable, resulting in a ranking of natural variables or a selection of characteristics. Small trees have an intuitive knowledge for experts in a given field because it is easy to extract rules from these trees simply by examining them. For deeper and wider trees, it is much more difficult to extract the rules and therefore the taller the tree is, the less intuitive its expression of knowledge. The smallest trees are obtained from the largest trees by size. Large trees often have high classification accuracy but not very good generalization capabilities. By reducing larger trees, we obtain smaller trees that often have better generalization capabilities (they avoid over-adjustment). Decision tree construction algorithms (e.g. C4.5) are relatively simpler than more complex algorithms such as SVMs.

### 2.4.3 SVM Support Vector Machines

The SVM is a classifier based on the search for a separation hyperplane in the feature space between two classes so that the distance between the hyperplane and the nearest data points in each class is maximized. The approach is based on a minimized classification risk [31] rather than an optimal classification. SVMs are well known for their generalizability and are particularly useful when the number of "m" characteristics is high and the number of learning data "n" is low ("m" "n") [4]. When the two classes are not separable, variables are added and a cost parameter is assigned to the overlapping data points. The maximum margin and place of the hyperplane are determined by quadratic optimization with a practical execution time of O(n 2), placing the SVM among the fast algorithms even when the number of attributes is high. Different types of divisional classification surfaces can be achieved by applying a nucleus, such as linear, polynomial, Gaussian Radial Base Function (GFR) or hyperbolic tangent. SVMs are binary classifiers, and multi-class classification is achieved by developing one SVM for each class pair. Hu et al. [25] used two robust support vector machines (RSVM), a variation of the SVM where the discriminant hyperplane is averaged to be smoother, and the regulation parameter is automatically determined, as the anomaly classifier in their study. The parts of the basic security module [32] of the 1998 DARPA dataset were used to pre-process the training and test data. The study showed good classification performance in the presence of noise (such as poor labeling of the training data set) and reported an accuracy of 75% without false alarms and an accuracy of 100%.

### 2.4.4 Fuzzy Logic

Fuzzy logic is derived from fuzzy set theory where reasoning is approximate rather than precisely deduced from classical predicate logic. Fuzzy techniques are therefore used in the field of anomaly detection mainly because the characteristics to be considered can be considered as fuzzy variables [19]. This type of treatment considers an observation to be normal if it is within a given interval [8]. Although fuzzy logic has proven effective, especially against scans and port probes, its main disadvantage is the high resource consumption involved. On the other hand, it should also be noted that fuzzy logic is controversial in some circles, and has been rejected by some engineers and most statisticians, who consider probability to be the only rigorous mathematical description of uncertainty.

### 2.4.5 Genetic Algorithms

Genetic algorithms are classified as global research heuristics and are a particular class of evolutionary algorithms that use techniques inspired by evolutionary biologies such as inheritance, mutation, selection, and recomposition. Thus, genetic algorithms are another type of technique based on automatic learning, capable of inferring classification rules [33] and selecting appropriate characteristics or optimal parameters for the detection process [19]. The main advantage of this subtype of learning is the use of a flexible and robust global search method that converges to a solution from multiple directions, while no prior knowledge of system behavior is assumed. Its main disadvantage is the high consumption of resources.

### 2.4.6 Clustering Construction

Clustering techniques work by grouping observed data into clusters, based on a given similarity or distance measurement. The most common procedure used is to select a representative point for each cluster. Then, each new data point is classified as belonging to a given cluster according to the proximity of the corresponding representative point [2]. Some points cannot belong to any cluster; they are called outliers and represent anomalies in the detection process. Clustering and outliers are currently used in the field of attack detection [34, 9], with several variations depending on how the question is distributed. For example, the KNN approach (nearest k-neighbor) [35] uses Euclidean distance to define the membership of data points in a given cluster, while others use Mahalanobis distance. Some detection proposals associate a certain degree of an outlier for each point. Clustering techniques determine the occurrence of attack events only from raw audit data, so the effort required to adjust the defense system is reduced.

The Multilayer Perceptron

The multilayer perceptron is a network of artificial neurons. It is composed of several layers of neurons, and the data is sent from one neuron in one layer to one or more neurons in the next layer [36]. Neurons perform operations on the input they receive using activation functions. The information is filtered many times, and the result is used to classify the algorithm data. Multilayer perceptrons have been studied for a long time and can be used to classify effectively, as shown by Atlas in an article in which it concludes that the performance of a trained multilayer perceptron equals or exceeds that of decision trees [37]. However, Belue has shown that it is essential to select the right characteristics of the algorithm data to obtain consistent results [38].

## 3 The Proposed Detection Method

The proposed framework for intrusion detection is a hybrid IDS. The proposed IDS uses the 6LoWPAN compression header based on the machine learning algorithms to learn and classify the type of attacks. Then, the rule or signature created by the machine learning algorithm is set to 6BR. Over time, when a new signature is available for routing attacks, 6BR will be updated with the new rule or signature generated by the new features. The proposed DIS framework is divided into three layers, as shown in Fig. 4. The first layer consists of detection agents that capture network data using the Cooja traffic analyzer. The captured data is then analyzed and filtered by a second layer model that extracts only the distinct characteristics that distinguish normal and abnormal network activities. At this level, the data is classified as normal or malicious (hello flood, wormhole or sinkhole). The proposed framework for the IDS is illustrated in Fig. 4.

### *3.1 Module 1: Dataset Generation*

The various RPL network communication scenarios were simulated by Cooja simulator. Cooja used the Contiki operating system [39]. The sensor nodes in the network implement the Routing Protocol for Low-Power and Lossy Networks (RPL) protocol

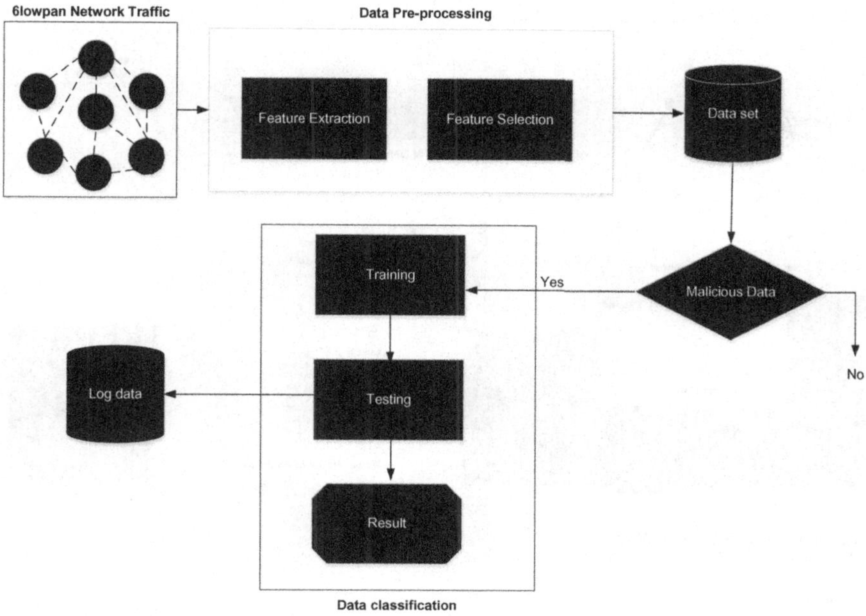


**Fig. 4** The proposed architecture

[34]. Contiki makes it possible to load and unload individual programs and services to the simulated sensors [40]. Figure 5 shows a Cooja User Interface. To simulate routing attacks such as hello-flood attack, wormhole attack, and sinkhole attack, we conducted a simulation scenarios of each attack on a large number of IoT nodes, up to 500, with different percentages (10, 20%, etc.) of malicious sensor nodes as shown in Table 1. We therefore simulated different routing attack scenarios and processed raw data sets to prepare them for the detection process. Subsequently, we used a Wireshark packet analyzer to transform OCAP files into CSV files. Then, we applied a data pre-processing script to extract the features of the generated CSV files. Finally, we concatenate the same attack data sets to obtain a complete data set to use in our research.

## 3.2 *Module 2: Data Pre-processing*

**A. Feature Extraction**

Once the scenarios are simulated, the data sets are produced as OCAP files. These files have been decomposed into CSV files using Wireshark. Machine learning algorithms require specific attributes of the learning data that are obtained by extracting characteristics. The data pre-processing step consists in extracting the relevant characteristics of the data to avoid calculation overload and to obtain problem-oriented

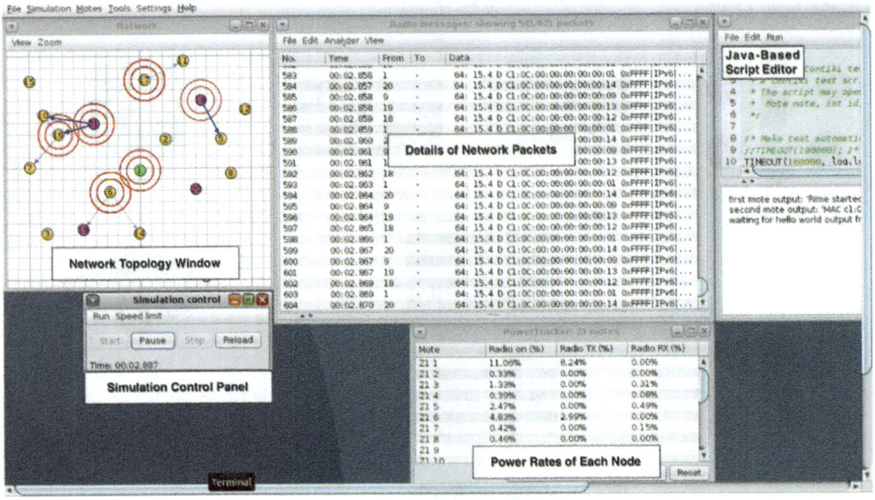


**Fig. 5** Cooja user interface

**Table 1** Datasets scenarios

| | Malicious | | | | Normal | | |
|---|---|---|---|---|---|---|---|
| Datasets | Scenarios | NB. nodes | NB malicious/ Normal nodes | Total NB packet | Scenarios | NB nodes | Total NB count |
| Hello flood | HF_10 | 10 | 2/8 | 212.134 | Normal_10 | 10 | 176.286 |
| | HF_50 | 50 | 8/50 | 328.465 | Normal_50 | 50 | 218.724 |
| | HF_100 | 100 | 16/84 | 416.274 | Normal_100 | 100 | 310.187 |
| | HF_500 | 500 | 50/500 | 675.765 | Normal_500 | 500 | 501.075 |
| Wormhole | WH_10 | 10 | 2/8 | 121.126 | Normal_10 | 10 | 118.172 |
| | WH_50 | 50 | 8/50 | 147.465 | Normal_50 | 50 | 129.557 |
| | WH_100 | 100 | 16/84 | 317.673 | Normal_100 | 100 | 238.933 |
| | WH_500 | 500 | 50/500 | 719.764 | Normal_500 | 500 | 450.193 |
| Sinkhole | SK_10 | 10 | 2/8 | 121.361 | Normal_10 | 10 | 117.213 |
| | SK_50 | 50 | 8/50 | 227.186 | Normal_50 | 50 | 165.763 |
| | SK_100 | 100 | 16/84 | 301.392 | Normal_100 | 100 | 216.031 |
| | SK_500 | 500 | 50/500 | 815.534 | Normal_500 | 500 | 721.554 |

attributes. We simulated different scenarios that have different topologies and network sizes for each type of attack. In the simulation result, we obtain the raw data sets. Cooja exports the PCAP and CSV files lafter the simulation is completed. However, raw data files are not sufficient to be the entry into the learning algorithm because the raw data set includes information such as source/destination node address and packet length, which causes noise and overadaptation in the learning algorithm.

For this reason, we have developed a feature extraction algorithm with Python 3.7. These libraries facilitate the mathematical operations necessary to extract the characteristics. We have set up a dictionary structure to handle a large number of nodes. We have chosen not to calculate global statistics on the total simulated time or a total number of packets, as this type of calculation could reduce the importance of the main characteristics extracted. Therefore, we divided all the simulations into periods, or windows of 5000 ms duration. Before this process, it is necessary to sort the data sets by simulation time, because the sequence of packet simulation time is significant for extracting characteristics and Cooja extracts PCAP files in the wrong time sequence. This is particularly true for a wide range of network topologies and long simulation times. The pseudocode of our pre-processing data algorithm is also indicated in the algorithm below.

**Data Extraction Algorithm:**

```
Input: pcap file
 getArguments pcap from network traffic;
Loop read each packet;
function
array ← dataset.csv
    Sorted array
    Feature conversion
Feature  Extraction:
            5000ms ← Window Size
            Calculate Feature values within the window size
            Labeling the dataset
End
End
```

Raw data sets have both quantitative and qualitative features. However, the learning algorithm used only accepts quantitative values. Therefore, we applied the conversion of characteristics to qualitative characteristics to transform their unified format. DAO (Destination Advertisement Object) is used in RPL for unicasting the training destination due to the selection of parents. DIO (Destination oriented directed acyclic graphs Information Object) is the most important type of message in RPL. It determines the best route through the base node using specific measures such as distance or countdown [41]. Another type of message is DIS. Nodes use DIS to join the network. Ack is a type of acknowledgment message to be used to give responses by nodes. Other types of messages in our data sets are the PDU (Protocol Data Unit) and UDP (Protocol Data Unit) packets, which are simulated data packets. The extracted features are listed in Table 2.

**Table 2** Extracted features

| Number | Abbreviation | Description |
|---|---|---|
| 1 | Num | Packet sequence number |
| 2 | Time | Time of simulation |
| 3 | Src | IP Source Node |
| 4 | Des | IP Destination Node |
| 5 | RT | Rate Transmission |
| 6 | RR | Rate Reception |
| 7 | ATT | Average Transmission Time |
| 8 | ART | Average Reception Time |
| 9 | PTC | Packet Transmitted Count |
| 10 | PRC | Packet Received Count |
| 11 | TTT | Total Transmission Time |
| 12 | TRT | Total Reception Time |
| 13 | DIO | DIO Packet Count |
| 14 | DAO | DAO Packet Count |
| 15 | DIS | DIS Packet Count |
| 16 | Tag | Malicious/Normal Label |

### *3.3 Module 3: Data Classification*

The data classification module configured the three routing attacks in the IoT at this layer, namely, hello flood, wormhole and sinkhole attacks. The essential attacks features are chosen for further analysis. Their packet number is tracked over time to study the behavior of each attack, and a set of rules is then developed. Then, the classes labeled "Normal," "Hello Flood," "Wormhole" and "Sinkhole" are created based on the revised rule. To classify each attack according to the defined classes, we compare six machine algorithms using the R language to find the most efficient algorithm. The algorithms are K-Nearest Neighbour (K-NN), Support Vector Machine (SVM), Naïve Bayes (NB), Random Forest (RF) and Multilayer Perceptron (MLP).

## 4 Implementation

During this step, the research plan is designed and can be implemented in practice. The whole implementation process can be outlined in the following steps:

1. Network traffic record (using Wireshark)
2. Feature extraction and selection (using Python 3.7)
3. Application of the machine learning methods (using R)
4. Evaluation of the results.

## 4.1 Evaluation Metrics

**Accuracy**: The accuracy of detection is measured as the percentage of correctly identified instances. This is the number of correct predictions divided by the total number of instances in the dataset. It should be noted that the accuracy is highly dependent on the threshold was chosen by the classifier and may, therefore, vary between different sets of tests. Therefore, this is not the optimal method to compare different classifiers, but it can give an overview of the class. Thus, the accuracy can be calculated using the following equation: The formula for calculating precision is given below (1):

$$Accuracy = \left(\frac{(\mathrm{TP} + \mathrm{TN})}{\mathrm{TP} \ + \mathrm{FP} + \mathrm{TN} + \mathrm{FN}}\right) \tag{1}$$

where:

| | |
|---|---|
| True positive (TP) | a number of positive samples correctly predicted. |
| False negative (FN) | number of positive samples wrongly predicted. |
| False positive (FP) | a number of negative samples wrongly predicted as positive. |
| True negative (TN) | number of negative samples correctly predicted |

**Precision**: Precision is defined as the proportion of true positive instances which are classified as positive. The precision tells that how many of the attacks are detected by model. The formula for calculating precision is given below (2):

$$Precision = TP/(TP + TN) \tag{2}$$

**Recall**: Recall, also commonly known as sensitivity, is the rate of the positive observations that are correctly predicted as positive. This measure is desirable, especially in the medical field because how many of the observations are correctly diagnosed. The sensitivity or the true positive rate (TPR) is defined by the formula below (3):

$$Recall = TP/(TP + FN) \tag{3}$$

**Energy Consumption**: Energy efficiency is an essential metric to adopt or not a security solution for constrained IoT applications. The evaluation of energy consumption is a key factor in estimating the lifetime of nodes. Equation (4) shows the energy usage per node, whereas equations below calculate the average of power consumed per second (5).

$$Energy\ (\mathrm{mJ}) = \begin{pmatrix} Transmit \times 19.5\,\mathrm{mA} + listen \times 21.8\,\mathrm{mA} + CPU \\ \times 1.8m1 + LPM \ \times 0.0545\,\mathrm{mA} \times (3\,\mathrm{V} \div 4096) \times 8 \end{pmatrix} \tag{4}$$

$$Power\,(\mathrm{mW}) = \frac{Energy\,(\mathrm{mJ})}{Time\,(\mathrm{s})} \tag{5}$$

**Memory Overhead**: RAM consumption is defined by statically pre-initialised and pre-zeroed variables, whereas ROM consumption is a size of an image loaded into a board. The obtained results refer to memory consumption for the whole Contiki image, which includes the entire communication stack. The total size of memory consumed in the experiment is calculated according to the equation below (6).

$$Total\ size = text + data + bss\ where\ bss\ is\ prezeroed\ RAM \tag{6}$$

## *4.2 Experimental Setup*

The three routing attacks in the IoT, namely, hello flood, sinkhole, and wormhole, are launched in Contiki's network simulator known as Cooja. Contiki has proven to be a powerful toolbox for building complex wireless systems and has shown a realistic result as in the real network [42, 31, 25]. Furthermore, all the data used in the simulation are from the real network environment. In the simulation, Tmote Sky is used as client node, and Cooja mote is used as a 6BR or sink node (Fig. 6).

One border router and one malicious hello flood/wormhole/sinkhole node. The border router is shown in green, the non-malicious nodes are shown in yellow and the malicious attack is shown in purple.

As shown in Table 1, we test the different ML models through 4 network setups (10 nodes, 50 nodes, 100 nodes and 500 nodes). In the experimentation part, we will present the results of our experiment with 500 nodes to show the effectiveness of the model proposed in large networks.

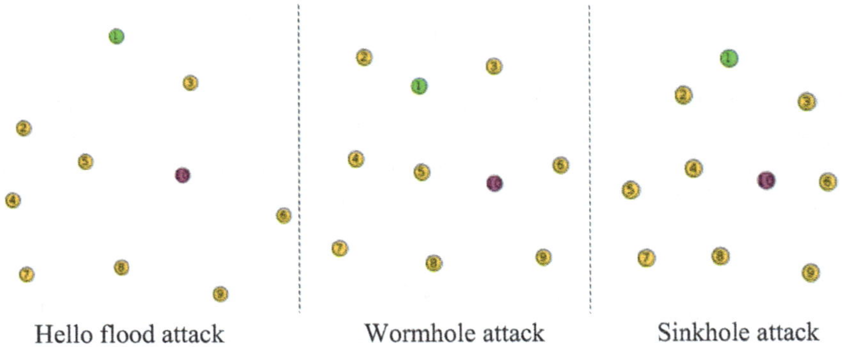


**Fig. 6** Network setup scenario with 10 nodes

### *4.3 Experimental Results and Evaluations*

From the experimental results, the best algorithm to classify routing attacks is obtained by analyzing five machine-learning algorithms, i.e., K-Nearest Neighbour (K-NN), Support Vector Machine (SVM), Naïve Bayes (NB), Random Forest (RF) and Multilayer Perceptron (MLP).

#### 4.3.1 Hello Flood Attack

We tested the different machine learning technique for Hello Flood Attack Detection Model with the proposed IoT intrusions dataset. The performance metrics are listed in Table 3.

The result shows that Random Forest and Naïve Bayes have the highest TP rate and can detect 100% of the Hello Flood attack as shown in Table 4. Random Forest reached 100% accuracy, recall and recall when tested on the Hello Flood data set. The K-NN algorithm is ranked second with a TP rate of 99.64% and an accuracy/precision of 99.97%.

**Table 3** Hello flood

| Classifiers | Evaluation criteria | | | | |
|---|---|---|---|---|---|
| | Precision (%) | Recall (%) | Accuracy (%) | TP rate (%) | FP rate (%) |
| K-NN | 99.7 | 99.3 | 99.7 | 99.64 | 0.021 |
| SVM | 97.7 | 97.6 | 97.7 | 93.50 | 0.063 |
| RF | 100 | 100 | 100 | 99.40 | 0.015 |
| NB | 97.1 | 96.3 | 96.3 | 97.80 | 0.027 |
| MLP | 98.9 | 98.9 | 98.9 | 96.85 | 0.022 |

**Table 4** Wormhole

| Classifiers | Evaluation criteria | | | | |
|---|---|---|---|---|---|
| | Precision (%) | Recall (%) | Accuracy (%) | TP rate (%) | FP rate (%) |
| K-NN | 99.6 | 99.3 | 99.6 | 97.36 | 0.048 |
| SVM | 99.7 | 98.8 | 99.7 | 95.77 | 0.072 |
| RF | 100 | 100 | 100 | 98.10 | 0.015 |
| NB | 100 | 100 | 100 | 97.34 | 0.018 |
| MLP | 100 | 100 | 100 | 97.25 | 0.015 |

### 4.3.2 Wormhole Attack

For the Wormhole attack Random Forest, Naïve Bayes and MLP achieve 100% detection, while other algorithms, SVM achieve 99.7% and K-NN achieve 99.6% as shown in Table 4.

### 4.3.3 Sinkhole Attack

In the Sinkhole attack, Random Forest again attains the best detection performances as shown in Table 5, with a 99.74% of precision, recall and accuracy and TP rate.

### 4.3.4 Energy and Memory Consumption

To measure the energy efficiency of our dataset with the proposed ML detection models. We compared the combined ML IDS with most popular public datasets that are preferred to be used in intrusion detection researches. UNSW-NB15 [43] and KDDCUP99 [44] are considered in this evaluation. The energy and memory consumption of each dataset with the combined ML techniques are compared to inspect their efficiency in a real environment implementation. Figure 7 shows the comparison of energy overhead for each IDS in a Tmote Sky node. The energy consumption of the proposed ML algorithms combined is 5840 mW.

Typically, constrained devices in IoT applications have limited memory. Thus, memory consumption is evaluated to assess the feasibility of IDS methods in constrained devices. In this assessment, the consumed memory of combined ML is 43.8 kB as shown in Fig. 8.

## 5 Conclusion and Future Works

IoT-6LoWPAN network nodes in aerospace cyber-physical systems are exposed to a variety of intrusion threats. IoT routing attacks (hello-flood attack, wormhole attack and sinkhole attack) are easily detected by the proposed ML detection models. This

**Table 5** Sinkhole

| Classifiers | Evaluation criteria | | | | |
|---|---|---|---|---|---|
| | Precision (%) | Recall (%) | Accuracy (%) | TP rate (%) | FP rate (%) |
| K-NN | 94.3 | 95.7 | 95.2 | 95.17 | 0.067 |
| SVM | 93.2 | 93.7 | 93.5 | 94.15 | 0.085 |
| RF | 99.5 | 99.6 | 99.7 | 99.74 | 0.004 |
| NB | 93.7 | 94.2 | 93.6 | 94.20 | 0.072 |
| MLP | 95.7 | 95.8 | 96.3 | 96.85 | 0.045 |

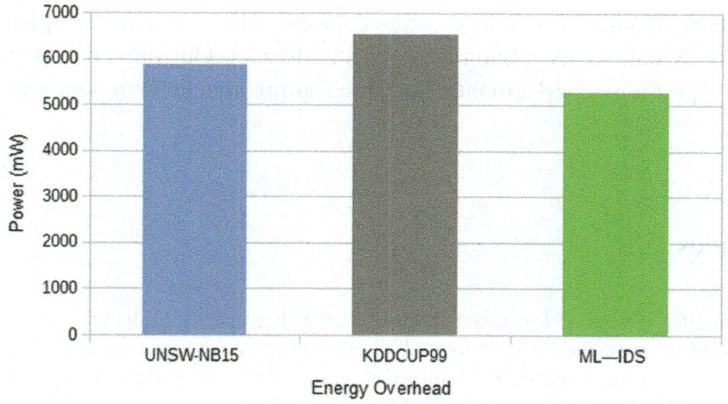


**Fig. 7** Energy overhead

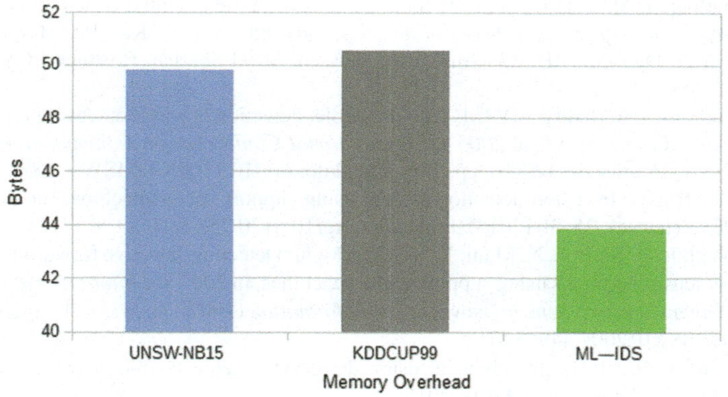


**Fig. 8** Memory overhead

chapter also fills a very important gap in the detection of routing attacks for IoT. The biggest challenge of this type of domains is the lack of data and existent datasets such as KDDcup 99 are too old. In this context, we generated real data from IoT-6LoWPAN network traffic recorded by network sniffers such as Wireshark. Also, we have built a detection model based on machine learning techniques, trained them with the routing attack data sets produced and created different attack detection models. We tested five machine learning algorithms: K-Nearest Neighbour (K-NN), Support Vector Machine (SVM), Naïve Bayes (NB), Random Forest (RF) and Multilayer Perceptron (MLP). Among these algorithms, the Random Forest algorithm is chosen for the proposed IDS because RF shows the best performance among all algorithms to detect routing attacks with 99% accuracy, precision and positive rate. The proposed IDS based on machine learning techniques effectively identified both individual and new anomaly attacks created by the combination of routing attacks.

For future work, further experiments will be conducted with many scenarios with more attacks and normal node rates to compare the effectiveness of the dataset used in this chapter with other intrusion detection datasets in the literature. We plan to enrich our IoT attack dataset by adding new routing attacks. Our objective is to increase the model's predictive performance for three routing attacks to include more routing attacks.

## References


1. L.A. Aguilar, The need for greater focus on the cybersecurity challenges facing small and midsize businesses. *Public Statement, US Securities and Exchange Commission* (2015)
2. R. von Solms, J. van Niekerk, From information security to cyber security. Comput. Secur. **38**, 97–102 (2013). http://dx.doi.org/10.1016/j.cose.2013.04.004
3. A. Plonk, A. Carblanc, Malicious software (malware): a security threat to the internet economy (2008)
4. T. Ramalingam, B. Christophe, F.W. Samuel, Assessing the potential of IoT in aerospace, in *Conference on e-Business, e-Services and e-Society*, ed. by A.K. Kar, P.V. Ilavarasan, M.P. Gupta, Y.K. Dwivedi, M. Mäntymäki, M. Janssen, S. Al-Sharhan (Springer, Cham, 2017), pp. 107–121
5. D. Janakiram, V.A. Reddy, A.V.U.P. Kumar, Outlier detection in wireless sensor networks using Bayesian belief networks, in *2006 1st International Conference on Communication Systems Software & Middleware* (2006), pp. 1–6. https://doi.org/10.1109/COMSWA.2006.1665221
6. J. Jha, L. Ragha, Intrusion detection system using support vector machine. Int. J. Appl. Inf. Syst. 2013 (Icwac), 25–30 (2013). https://doi.org/10.5120/758-993
7. S. Kaplantzis, A. Shilton, N. Mani, Y.A. Sekercioglu, Detecting selective forwarding attacks in wireless sensor networks using support vector machines, in *2007 3rd International Conference on Intelligent Sensors, Sensor Networks and Information* (2007), pp. 335–340. https://doi.org/10.1109/ISSNIP.2007.4496866
8. Y. Maleh, A. Ezzati, Lightweight intrusion detection scheme for wireless sensor networks. IAENG Int. J. Comput. Sci. **42**(4) (2015)
9. Y. Zhang, N. Meratnia, P.J.M. Havinga, Distributed online outlier detection in wireless sensor networks using ellipsoidal support vector machine. Ad Hoc Netw. **11**(3), 1062–1074 (2013). https://doi.org/10.1016/j.adhoc.2012.11.001
10. Y. Maleh, A. Ezzati, M. Belaissaoui, An enhanced DTLS protocol for Internet of Things applications, in *Proceedings—2016 International Conference on Wireless Networks and Mobile Communications, WINCOM 2016: Green Communications and Networking* (2016). https://doi.org/10.1109/WINCOM.2016.7777209
11. E.M. Atkins, J.M. Bradley, Aerospace cyber-physical systems education, in *AIAA Infotech@Aerospace (I@A) Conference*. American Institute of Aeronautics and Astronautics (2013). https://doi.org/10.2514/6.2013-4809
12. S. Berkovich, Physical world as an Internet of Things, in *Proceedings of the 2nd International Conference on Computing for Geospatial Research & Applications* (New York, NY, USA, ACM, 2011), pp. 66:1—66:2. https://doi.org/10.1145/1999320.1999389
13. G. Schuh, T. Potente, C. Thomas, A. Hauptvogel, Cyber-physical production management. *In IFIP International Conference on Advances in Production Management Systems*. 477–484 (2013, September). Springer, Berlin, Heidelberg.
14. J. Shi, J. Wan, H. Yan, H. Suo, A survey of cyber-physical systems. *In 2011 international conference on wireless communications and signal processing (WCSP)*.1–6. IEEE. (2011, November)
15. D. Strang, R. Anderl, Assembly process driven component data model in cyber physical production systems. *In Proceedings of the World Congress on Engineering and Computer Science*. **2**, (2014)

16. A. Humayed, J. Lin, F. Li, B. Luo, Cyber-physical systems security—a survey. IEEE Internet Things J. **4**(6), 1802–1831 (2017). https://doi.org/10.1109/JIOT.2017.2703172
17. H. Kim, Security and vulnerability of SCADA systems over ip-based wireless sensor networks. Int. J. Distrib. Sens. Netw. (2012). https://doi.org/10.1155/2012/268478
18. K. Ashton, That "Internet of Things" Thing. RFiD J. **22**(7), (2011)
19. Z. Shelby, C. Bormann, *6LoWPAN: The Wireless Embedded Internet—Shelby—Wiley Online Library* (Wiley, 2011)
20. G. Mulligan, The 6LoWPAN architecture. *In Proceedings of the ACM 4th Workshop on Embedded Networked Sensors*, 78–82 (2007)
21. Y. Maleh, A. Ezzati, M. Belaissaoui (eds.), *Security and Privacy in Smart Sensor Networks*. IGI Global (2018)
22. H.K. Patil, T.M. Chen, Wireless sensor network security, in *Computer and Information Security Handbook* (Elsevier, 2017), pp. 317–337. https://doi.org/10.1016/B978-0-12-803843-7.00018-1
23. H. Suo, J. Wan, C. Zou, J. Liu, Security in the Internet of Things: A Review. *In 2012 International Conference on Computer Science and Electronics Engineering*, **3**, 648–651 (2012).https://doi.org/10.1109/ICCSEE.2012.373
24. Z. Benenson, P.M. Cholewinski, F.C. Freiling, Vulnerabilities and attacks in wireless sensor networks, in *Wireless Sensor Network Security*, pp. 22–43 (2007)
25. S. Raza, L. Wallgren, T. Voigt, SVELTE: real-time intrusion detection in the Internet of Things. Ad Hoc Netw. **11**(8), 2661–2674 (2013). https://doi.org/10.1016/j.adhoc.2013.04.014
26. R.J. Cai, X.J. Li, P.H.J. Chong, A novel self-checking trad ad hoc routing scheme against active black hole attacks. Secur. Commun. Netw. **9**(10), 943–957 (2016). https://doi.org/10.1002/sec.1390
27. S.A. Kumar, T. Vealey, H. Srivastava, Security in Internet of Things: challenges, solutions and future directions, in *2016 49th Hawaii International Conference on System Sciences (HICSS)* (2016), pp. 5772–5781. https://doi.org/10.1109/HICSS.2016.714
28. Y. Maleh, A. Ezzati, M. Belaissaoui, DoS attacks analysis and improvement in DTLS protocol for Internet of Things. Proc. Int. Conf. Big Data Adv. Wirel. Technol. **54**(1–54), 7 (2016). https://doi.org/10.1145/3010089.3010139
29. C. Perkins, E. Belding-Royer, S. Das, *Ad hoc on-demand distance vector (AODV) routing*. RFC 3561 (2003)
30. C.E. Perkins, P. Bhagwat, Highly dynamic destination-sequenced distance-vector routing (DSDV) for mobile computers. ACM SIGCOMM Comput. Commun. Rev. **24**(4), 234–244 (1994). https://doi.org/10.1145/190809.190336
31. P. Pongle, G. Chavan, Real time intrusion and wormhole attack detection in Internet of Things. Int. J. Comput. Appl. **121**(9) (2015)
32. Z. Shelby, C. Bormann, *6LoWPAN: The Wireless Embedded Internet. 6LoWPAN: The Wireless Embedded Internet* (2009). https://doi.org/10.1002/9780470686218
33. S. Thirumuruganathan, A detailed introduction to K-nearest neighbor (KNN) algorithm. WWW Document (2010). https://saravananthirumuruganathan.wordpress.com/2010/05/17/a-Detailed-Introduction-to-K-Nearest-Neighbor-Knn-Algorithm/
34. T. Winter, P. Thubert, A. Brandt, T.H. Clausen, J.W. Hui, R. Kelsey, J. Vasseur, Rpl: Ipv6 routing protocol for low power and lossy networks (2011). Http://tools. Ietf. Org/html/draft-Ietf-Roll-Rpl-19, (July), 1–164. https://doi.org/10.2313/NET-2011-07-1
35. S. Thirumuruganathan, A Detailed Introduction to K-Nearest Neighbor (KNN) Algorithm. (2010). WWW Document. Available at: https://Saravananthirumuruganathan.Wordpress.Com/2010/05/17/a-Detailed-Introduction-to-k-Nearest-Neighbor-Knn-Algorithm/.
36. S.K. Pal, S. Mitra, Multilayer perceptron, fuzzy sets, and classification. IEEE Trans. Neural Networks. **3**(5), 683–697 (1992)
37. L. Atlas, R. Cole, Y. Muthusamy, A. Lippman, J. Connor, D. Park, R.J. Marks, A performance comparison of trained multilayer perceptrons and trained classification trees. *Proceedings of the IEEE* (1990), **78**(10), 1614–1619. https://doi.org/10.1109/5.58347

38. M. Mathews, M. Song, S. Shetty, R. Mckenzie, Detecting compromised nodes in wireless sensor networks (2007), pp. 273–278. https://doi.org/10.1109/SNPD.2007.538
39. L.M. Belue, K.W. Bauer, Determining input features for multilayer perceptrons. Neurocomputing. **7**(2), 111–121 (1995). https://doi.org/10.1016/0925-2312(94)E0053-T
40. B.A. Bagula, Z. Erasmus, Iot Emulation with Cooja, (March), 1–44 (2015). Retrieved from http://wireless.ictp.it/school_2015/presentations/firstweek/ICTP-Cooja-Presentation-version0.pdf
41. A. Dunkels, B. Gronvall, T. Voigt, Contiki—a lightweight and flexible operating system for tiny networked sensors, in *29th Annual IEEE International Conference on Local Computer Networks* (2004), pp. 455–462. https://doi.org/10.1109/LCN.2004.38
42. A. Al-Fuqaha, M. Guizani, M. Mohammadi, M. Aledhari, M. Ayyash, Internet of Things: a survey on enabling technologies, protocols, and applications. IEEE Commun. Surv. Tutor. **17**(4), 2347–2376 (2015). https://doi.org/10.1109/COMST.2015.2444095
43. M.N. Napiah, M.Y.I. Bin Idris, R. Ramli, I. Ahmedy, Compression header analyzer intrusion detection system (CHA—IDS) for 6LoWPAN communication protocol. IEEE Access **6**, 16623–16638 (2018). https://doi.org/10.1109/ACCESS.2018.2798626
44. N. Moustafa, J. Slay, UNSW-NB15: a comprehensive data set for network intrusion detection systems (UNSW-NB15 network data set), in *Military Communications and Information Systems Conference (MilCIS)* (2015), pp. 1–6
45. M. Tavallaee, E. Bagheri, W. Lu, A.A. Ghorbani, A detailed analysis of the KDD CUP 99 data set, in *2009 IEEE Symposium on Computational Intelligence for Security and Defense Applications* (2009), pp. 1–6. https://doi.org/10.1109/CISDA.2009.5356528

The manufacturer's authorised representative in the EU is Springer Nature Customer Service Centre GmbH, Europaplatz 3, 69115 Heidelberg, Germany. If you have any concerns regarding our products, please contact ProductSafety@springernature.com

Printed and bound by CPI Group (UK) Ltd, Croydon, CR0 4YY
30/09/2026
02225495-0007